Philip H. Gosse, Charles T. Hudson

The Rotifera

or Wheel-Animalcules - Vol. 1

Philip H. Gosse, Charles T. Hudson

The Rotifera
or Wheel-Animalcules - Vol. 1

ISBN/EAN: 9783337238841

Printed in Europe, USA, Canada, Australia, Japan

Cover: Foto ©berggeist007 / pixelio.de

More available books at **www.hansebooks.com**

THE ROTIFERA;

OR

WHEEL-ANIMALCULES.

BY

C. T. HUDSON, LL.D. Cantab.

ASSISTED BY

P. H. GOSSE, F.R.S.

IN TWO VOLUMES——VOLUME I.

WITH ILLUSTRATIONS.

Throughout the whole of the work Dr. Hudson has had the invaluable assistance of Mr. Gosse's MS. notes, and of his close and constant revision of the proofs.

The hearty thanks of the authors are due to Mr. Frank Crisp, one of the secretaries of the Royal Microscopical Society, and editor of its Journal, for the great service that he has rendered them by forwarding early notices of all the pamphlets and papers published on the subject : anyone who is, or has been, engaged in a task similar to this will know what time and labour have thus been spared, by his thoughtful kindness.

They are, too, greatly indebted to numerous kind correspondents for living specimens, often obtained with no little expenditure of time and trouble, and for the records of their observations on them : as well as for valuable preparations of rare species, which could not be sent alive : to these correspondents the authors tender their grateful thanks. In each case where the specimen was new or rare the name of the place in which it was found, and (when permitted) the name also of the finder, has been added to the description. They wish, however, to make special mention of how much they owe to Miss Saunders of Cheltenham, Miss Davies of Woolston, Dr. F. Collins, and Mr. John Hood of Dundee, for their constant kindness in sending a profuse number of specimens of many species, some of which were of unusual interest.

It is so natural to recommend one's own favourite pursuit that the recommendation often carries but little weight ; and yet there is much to be said in favour of the study of the Rotifera, that cannot be gainsaid. They are to be found almost everywhere ; they cost nothing ; they require neither expensive lenses nor an elaborate apparatus ; they tempt us to explore the country, and to take pleasant walks ; they are beautiful themselves ; and they suggest all kinds of difficult questions on life and being. Moreover there is happily still a great store of scientific ignorance concerning them, thus leaving an ample field for fresh discoveries.

Nor is this all. The study of these animated specks (in which teeth, stomach, muscles, and even a brain lie hidden in the compass of an invisible mote) irresistibly leads the mind to the contemplation of Him, whose almighty hand is as visible in an atom of this animated dust, as it is in the myriad sparkles of the starlit heavens.

C. T. H.

CONTENTS

OF

THE FIRST VOLUME.

—◦◦<◦>◦◦——

THE ROTIFERA.

CHAPTER I.

INTRODUCTION.

Contemplatio Naturæ præegustus est voluptatis cælestis, constans animi gaudium, perfectique ejus solatii initium, summusque felicitatis humanæ apex. Cum Anima hujus particeps fuerit, ex gravi quasi sopore excitata, in luce ambulat, sui ipsius obliviscens, in cælesti, ut ita dicam, terra, inque terrestri cælo.—J. BASTER.

Inest in explicatione Naturæ, insatiabilis quædam e cognoscendis rebus Voluptas, in qua una, confectis rebus necessariis, vacui negotiis, honeste ac liberaliter possumus vivere. — CICERO, *De Finibus*, lib. iv. cap. 5.

CHAPTER I.

INTRODUCTION.

On the Somersetshire side of the Avon, and not far from Clifton, is a little combe, at the bottom of which lies an old fish-pond.

Its slopes are covered with plantations of beech and fir, so as to shelter the pond on three sides, and yet leave it open to the soft south-western breezes, and to the afternoon sun. At the head of the combe wells up a clear spring, which sends a thread of water, trickling through a bed of osiers, into the upper end of the pond. A stout stone wall has been drawn across the combe from side to side, so as to dam up the stream; and there is a gap in one corner, through which the overflow finds its way, in a miniature cascade, down into the lower plantation.

The pond's smooth surface is prettily diapered with the green leaves of many a water-plant, and with the sharp images of three famous beeches growing close to its edge: but to a naturalist's eye the old wall is the more charming object. Time has crumbled away the mortar near the water's edge, and made a thousand nooks and crannies; which, densely clothed with algæ, are the haunts of myriads of living creatures.

If we approach the pond by the gamekeeper's path from the cottage above, we shall pass through the plantation, and come unseen right on to the corner of the wall; so that one quiet step will enable us to see at a glance its whole surface, without disturbing any living thing that may be there.

Far off at the upper end a water hen is leading her little brood among the willows; on the fallen trunk of an old beech, lying half-way across the pond, a vole is sitting erect, rubbing his right ear; and the splash of a beech husk just at our feet tells of a squirrel, who is dining somewhere in the leafy crown above us.

But see! the water rat has spied us out, and is making straight for his hole in the bank, while the ripple above him is the only thing that tells of his silent flight. The water hen has long ago got under cover, and the squirrel drops no more husks. It is a true 'Silent Pool,' and without a sign of life.

But if, retaining sense and sight, we could shrink into living atoms and plunge under the water, of what a world of wonders should we then form part! We should find this fairy kingdom peopled with the strangest creatures:—creatures that swim with their hair, that have ruby eyes blazing deep in their necks, with telescopic limbs that now are withdrawn wholly within their bodies and now stretched out to many times their own length. Here are some riding at anchor, moored by delicate threads spun out from their toes; and there are others flashing by in glass armour, bristling with sharp spikes or ornamented with bosses and flowing curves; while, fastened to a green stem, is an animal convolvulus that by some invisible power draws a never-ceasing stream of victims into its gaping cup, and tears them to death with hooked jaws deep down within its body.

Close by it, on the same stem, is something that looks like a filmy heart's-ease. A curious wheelwork runs round its four outspread petals; and a chain of minute things, living and dead, is winding in and out of their curves into a gulf at the back of the

B 2

flower. What happens to them there we cannot see; for round the stem is raised a tube of golden-brown balls, all regularly piled on each other. Some creature dashes by, and like a flash the flower vanishes within its tube.

We sink still lower, and now see on the bottom slow-gliding lumps of jelly that thrust a shapeless arm out where they will, and, grasping their prey with these chance limbs, wrap themselves round their food to get a meal; for they creep without feet, seize without hands, eat without mouths, and digest without stomachs.

Time and space, however, would fail me to tell of all the marvels of the world beneath the waters. They would sound like the wild fancies of a child's fairy tale, and yet they are all literally true; and, moreover, nearly all of them are true of that rotiferous world which it is my purpose to describe.

But it will be naturally asked by those of my readers to whom the subject is new, " What *is* a Rotiferon ? " and no doubt one would say that a book about Rotifera ought to begin at the beginning, and define precisely what a Rotiferon is.

Precise definition is, however, in such a case, quite out of the question; for, though it is easy enough to define the typical form of a natural group of animals, or even to include in the definition forms that must be placed not far off from the central one, yet in the ambitious attempt to frame a definition that shall include many families, we find (as we get farther away from the typical form) that one by one all the positive statements are disappearing from our definition; and at last we have nothing left but the mere shell of a proposition, with everything worth the stating struck out of it.

The Rotifera, then, are small aquatic animals varying from $\frac{1}{8}$ to $\frac{1}{300}$ of an inch in length, and deriving their name from a wheel-like appearance produced by fine circlets of hairs seated on the front of their heads. A few species are marine; but the great majority known to us belong to fresh water, and are to be found in ditches, ponds, reservoirs, lakes, and slowly running streams, sometimes attached to the leaves and stems of water plants, sometimes creeping on the algæ, sometimes swimming freely through the water. Although the greater number of the genera resemble each other in the chief features of their internal organization, so as to form a very natural group of animals, yet there are several aberrant forms which would render it a difficult matter to include them all in one precise definition.

This indeed could be done only by introducing so many qualifications and exceptions to every statement, that the portrait would be rendered too vague for any reader but one already familiar with the whole subject.

Of the greater number, however, it is enough to say :

(1) That they swim by means of hairs on the front of their heads.

(2) That they possess a simple stomach and intestine : and peculiar jaws.

(3) That they have muscles which are sometimes striated, and which often pass freely through the cavity of the body.

(4) That they have a well-developed vascular system.

(5) That their nervous system consists of one ganglion, with nerve threads radiating to their organs of sense.

(6) That they are diœcious; have ova of two kinds; and do not pass through any distinct metamorphosis.

Though the above six statements are precise enough, and in the main true, yet it will be as well for those, who are not versed in the subject, to pass them over for the present, and first to master the structure of some one typical Rotiferon; as, when this has been done, the general conception of a Rotiferon will be easily grasped, and the variations from the type readily followed and understood.

For this purpose I have selected *Brachionus rubens*, whose figure is given much magnified in Pl. A, fig. 1. The genus *Brachionus* is to be met with almost everywhere. It is hardly possible in summer to take a dip of water from a garden-pond, or to gather the algæ from its walls, without bringing up some specimen of the genus. *Brachionus rubens* is a fairly common species. It is comparatively a large handsome animal, very suitable for the purpose of description, and one which bears the temporary captivity of a compressorium remarkably well.

The Female.

Fig. 1, Pl. A, represents the dorsal aspect of the female of this *Brachionus*, and fig. 2 the upper part of the ventral aspect. The drawings are from life; but the outlines of the various organs have been made unnaturally sharp and distinct, for the sake of clearness. The dorsal and ventral surfaces may be distinguished from each other in the great majority of the Rotifera by the following considerations:

First, as to the dorsal surface:

(1) It is arched (fig. 5).

(2) The stomach (fig. 1, *s*) passes down it; between it and the ovary (fig. 1, *oy*).

(3) The cloaca (fig. 1, *cl*) is on it; in the median line.

(4) There is almost invariably one antenna (fig. 1, *a*) (or a coalesced pair) on it; placed anteriorly on the median line.

(5) The eye or eyes (fig. 1, *e*) are towards the dorsal surface.

(6) In swimming over objects the Rotiferon keeps the dorsal side upwards.

Secondly, as to the ventral surface (fig. 2):

(1) It is comparatively flat.

(2) The entrance to the mouth lies on it (figs. 2, 5, *bf*).

(3) The ovary is placed close to it (fig. 2, 5, *oy*).

In the case of those Rotifera whose dorsal and ventral surfaces have much the same contour, the above considerations present points of difference enough to decide between the two.[1]

B. rubens is inclosed in a case or *lorica* (figs. 3, 4, 5) which is both hard and transparent. The internal structure can be readily seen through it; and, by suffering the animal to dry on a glass slip, and then dropping on it a solution of caustic potash, the softer portions of the body may be dissolved away, and the lorica left unharmed.

It will then be seen to be closed above and below, with an opening at each end, like the shell of a tortoise. From the front opening the head is protruded, and from the hind the *pseudopodium* or *foot* (fig. 1, *f*).

The lorica has a glassy shining surface, and is armed with six short sharp spines in front, of which the central pair is the longest. Four of them are distinctly on the dorsal surface; but the outmost pair belongs as it were to both surfaces, being on the edge where they meet.

The front edge of the ventral surface (the *mental* edge as it is sometimes termed)[2] is hollowed out symmetrically into graceful curves (fig. 4).

The lorica widens from the front backwards, till, at about two-thirds of its length, it reaches its maximum breadth; and is then rounded off by two ogee curves that are separated, by a square notch on the dorsal surface, but by a nearly circular one on the ventral. In consequence, it is often said that the lorica has two blunt spines behind on the dorsal surface; but this is somewhat misleading, as these so-called spines are merely the sides of the excavation. Strong ridges from each of the four central front spines run down about one-third of the dorsal surface (fig. 3), and still longer ridges mark the ventral surface with sweeping curves (fig. 4).

The median portion of the lorica is by far the deepest, and in it the internal organs mainly lie. The dorsal surface of the lorica slopes upwards from the head to its line of greatest width (fig. 5), and then abruptly falls to meet the under surface; the whole lorica thinning off there into closely approaching plates, through the excavation in which the foot can pass. Each side, too, of the median portion of the lorica thins off in a similar manner; so that the dorsal and ventral surfaces meet everywhere (except at the head) in a sharp edge.

[1] *Cf.* Dr. Moxon, *Trans. Linn. Soc.* vol. xxiv. 1864, p. 455.   [2] From *mentum*, the chin.

The head is shaped somewhat like a truncated cone, with the larger end forward; posteriorly it is studded with several small rounded lobes; while from its anterior surface rise three fleshy protuberances, crowned with stout vibrating hairs called *styles*.

Each side of the front of the head, or *corona*,[1] is rounded into a nearly circular lobe, and along the rim of each lobe runs an unbroken row of smaller vibrating hairs, called *cilia*, which are continued so as to meet each other on the dorsal surface. It is by means of this apparatus that *Brachionus* both swims and procures its food.

As the head is seldom withdrawn into the lorica for more than a moment or two, and as the cilia begin to play the instant it is protruded, *Brachionus* would have been condemned to almost perpetual motion if it had not been for the foot. This organ is provided at its extremity with two pincer-like processes, or *toes* (fig. 1, *t*); which, however, do not pinch, but which can adhere even to glass by means of a viscid secretion that flows through their tips.

The *foot-glands* (fig. 1, *fg*), which secrete this substance, are two club-shaped organs running down the whole length of the foot. They are to be met with in nearly all the Rotifera.

The cilia, which are set closely round the edge of the corona, lash the water with such fury that it is impossible usually to follow the action of any individual cilium; but, by selecting an animal whose corona is close to the covering glass of the live box, some spot can often be found where the action of the cilia is checked by their striking against the glass; and, under these circumstances, it is easy to understand their action. Each cilium lashes sharply downwards (like a whip) on the corona, and then rises gradually into its place again, to repeat the action continuously, so long as the corona is expanded. As, however, the cilia do not do this simultaneously, but in turn, one after the other, in very rapid succession, those that can be seen together at any given moment are in every phase, from complete extension to complete depression; thus giving rise to various wave-like illusory appearances, according to the illumination, and also to the plane on which the objective is brought to focus. One of the most common of these appearances is that of a toothed wheel, which is so well imitated by the *Philodinadæ*, that early observers thought such wheels existed, and drew them like the escapement-wheel of a watch.[2]

If a little carmine be mixed with the water, two beautiful coloured spirals will be produced by the action of the ciliary wreaths, one on each side of the head leading down to the buccal funnel. The orifice of the buccal funnel, or, as it will be termed, the *buccal orifice*, lies in a niche on the ventral surface; it is fringed by the ciliary wreath, which here dips down on either side of the corona, and passes round the V-shaped opening of the *buccal funnel* (fig. 2, *bf*)—that is, of the passage leading from the niche to the mastax. The atoms brought by the ciliary currents pass down the buccal funnel, which itself is lined with cilia; and, if uninterrupted, enter the *mastax* (figs. 1, 2, *mx*), a muscular bulb containing the *trophi* (fig. 1, *ti*) or teeth. But it is not every atom whirled down the buccal funnel that is suffered to reach the mastax; for there are two lip-like processes (fig. 2, *lp*) rising from the mastax, which can be seen every now and then thrust up and down the buccal funnel; and which by closing prevent the passage of morsels that are not to the Rotiferon's taste. The sudden check, produced by the lips on the inflowing current, always sweeps out of the buccal funnel whatever the animal desires to reject; and a constant stream of rejected particles may be seen issuing from the buccal funnel midway between the spirals caused by the corona.

The Mastax.

The *mastax* (figs. 1, 2, *mx*; fig. 6) is the muscular covering of the jaws or *trophi*. It has thick walls, and is slightly three-lobed, each lobe investing one of the three principal parts of which the trophi consist. There is an opening in front towards the ventral surface at the bottom of the buccal funnel, whose walls here merge into those of

[1] A name suggested by Mr. Cubitt in lieu of *trochal disk*.
[2] *Baker on the Microscope*, vol. i. Plate VIII. fig. 6, 1785.

the mastax; and there is a posterior opening towards the dorsal surface, just above the œsophagus.

The contained hard parts, or *trophi*, consist of two hammer-like bodies, the *mallei* (fig. 6, *ms*; fig. 9), and of a third anvil-piece called the *incus* (fig. 6, *is*). Each malleus has for its head or *uncus* (figs. 6, 7, 9, *us*) a piece which, when spread out by pressure, is like a comb with five unequal teeth (fig. 8, *us*), but which, under ordinary circumstances, is much curved, so as to bring the teeth close together.

The handle of each of these hammers is a single stout piece called the *manubrium* (figs. 6, 7, 8, 9, *mm*). The anvil or *incus* roughly resembles a triangular prism, of which one end (the dorsal one) has been tapered to a point (fig. 10). It is divided into two equal pieces, the *rami* (figs. 6, 7, *rs*), by a cut from the dorsal to the ventral surface, so as to leave them just attached to each other at the bottom of the broader end. The two rami rest there on what looks like a stem, the *fulcrum* (figs. 6, 7, *fm*), but which viewed laterally is seen to be a slender plate (fig. 10, *fm*).

Muscles, springing from the walls of the mastax, are attached to various parts of the mallei and rami, and act so as to cause the unci to approach and recede from each other. But each uncus is fastened to the corresponding ramus; and, in consequence, as the mallei recede the rami are opened, and when the mallei approach each other the rami are closed.[1]

From these arrangements it results that all the food which falls from the buccal funnel into the mastax, is first torn by the sharp points of the unci and then crushed between the opposing surfaces of the rami. On passing the rami, the food enters at once into the œsophagus.

The Œsophagus.

The *œsophagus* (fig. 2, *œ*) is a short tube, with thick walls, connecting the mastax with the stomach. The thickness of the walls may be readily seen by feeding *Brachionus* with carmine, and watching the stream of coloured particles trickle through the œsophagus into the stomach. Their course, in a narrow central line, shows clearly the small dimensions of the passage.

There is always an appearance of water flowing through the œsophagus, even when the animal has withdrawn its head into the lorica, and when of course the action of the ciliary wreath has ceased. This appearance seems to be due to minute cilia lining the œsophagus, which by their joint rhythmical action give rise to illusory waves much larger than any real waves could be. That this explanation is not a mere guess will be shown farther on in the case of the Floscules.

The Glands.

Seated near the top of the mastax, and on either side of the buccal funnel, are two clear vesicular organs; which are possibly *salivary glands* (fig. 2, *sg*), and are unusually large in *Brachionus*. They consist of two or three lobes on either side, and are best seen from the ventral side, as the buccal funnel enters the mastax, a little below the summit, on that side.

There is a pair of glands called the *gastric glands* (figs. 1, 2, *gg*) placed at the anterior end of the stomach, and on either side of it. They exhibit cells with central nuclei imbedded in a granular substance. Each has a long, wide, nucleated duct connecting it with the upper end of the stomach. They probably perform the function of a rudimentary liver, and possibly cause, by their secretion, the peculiar yellow-brown colour which is so frequently met with in the stomach. The *foot-glands* have already been described in p. 6.

The Stomach and Intestine.

The stomach (figs. 1, 2, *s*) is a simple sack, with thick cellular walls, which are frequently

[1] For a full description of the complex action of the trophi see Mr. Gosse's paper "On the Structure, Functions, and Homologies of the Manducatory Organs in the class Rotifera," *Phil. Trans.* 1856, p. 419.

studded with what appear to be bright yellow oil-globules. It is divided by an in-vagination at its lower third into an upper portion, the true stomach, and a lower which may be considered to be a short intestine (fig. 1, *i*). This lower portion frequently lies transversely to the longer axis of the stomach. Both are thickly ciliated on the inner surface, but the cilia of the intestine are larger, and more readily seen.

When a portion of digested food has been transferred from the stomach to the in-testine, it is kept slowly revolving by the cilia, till it is suddenly expelled through the cloaca (fig. 1, *cl*). The intestine is connected with the cloaca by a short and very dilatable tube or *rectum* (fig. 1, *r*), and ends (as has been already said) on the dorsal surface, in the median line, just at the commencement of the foot. The rectum also is ciliated, so that the whole of the alimentary tract from the top of the buccal funnel down to the cloaca, with the exception perhaps of the passage through the mastax, is lined with cilia.

The Vascular System.

At the right of the intestine (viewed dorsally), and just under the line of the lorica's greatest width, lies the *contractile vesicle* (fig. 1, *cv*). This is a delicate bladder which alternately dilates and contracts, and with some regularity.

The contraction is produced by fine muscular threads, which ramify in its walls, and cause it to empty its contents through a duct into the cloaca. Its distension is most probably due to the fluid poured into it by two looped and twisted tubes (fig. 1, *lc*), which may be seen passing to it, one on each side of the body down from the head.

This is, however, a much disputed question, which will be discussed fully in another place, along with the probable function of the whole apparatus.

The tubes appear to be surrounded with a granular flocccose material, which here and there dilates into irregular masses. Attached to the tubes on each side, at tolerably regular intervals, are five little tag-like bodies (fig. 1, *vt*), in which a flickering motion may be constantly seen, sometimes presenting the appearance of a waving cilium. There is much difference of opinion about the true structure of these tags—the *vibratile tags*, as they are termed—but it is probable that their office is to direct the periviscceral fluid into the tubes, and along them into the contractile vesicle, whence it is driven at intervals through the cloaca.

The Muscles.

The dorsal muscles are shown in fig. 3, and the ventral in fig. 4. From the posterior dorsal surface of the head, on each side of the cephalic ganglion, and close to it, a stout muscle (fig. 3, 1, 1) slopes backward towards the dorsal surface, and is attached by a broad base to the lining membrane of the lorica. Outside this pair is a second (fig. 3, 2, 2), similarly attached, and running rather obliquely underneath the first pair, but not quite so stout. A similar pair (fig. 1, 4, 4) is attached to the posterior ventral surface of the head, and to the lining of the ventral surface of the lorica. The united action of these three pairs of muscles withdraws the head into the lorica.

When it is so withdrawn, a pair of diverging muscular threads (not given in the figure) can be seen fixed to the lorica, just below its central notch, with their other ends fastened to the head. These evidently oppose the action of the three other pairs (figs. 3, 1, 1, 2, 2; 4, 4, 4) and help to draw out the head again. They are assisted in this by a further pair of muscles (also omitted from figs. 3 and 4), each of which is fastened at one end to the base of one of the outermost anterior spines, and at the other to a side lobe of the head.

But the principal part in driving out both the head and the foot is borne by transverse muscles, which are attached to the lorica at the side, and are closely applied through-out their length to the soft organs of the body. Their sudden contraction compresses the periviscceral fluid, and so forces out the retracted head or foot. Nothing could be

more effective than this hydrostatic pressure ; and under it the retracted foot darts out
of the lorica with amazing swiftness.

When the head is protruded, and the cilia are all in full play, *Brachionus* may often
be seen to move its head, without withdrawing it, first to one side and then to the other,
depressing the side lobes alternately. This action is effected by two pairs of short
muscles (figs. 3, 4, 3, 3), one on each side, attached to the lorica at about one-third of its
length from the front, and at their anterior ends inserted into the side lobes.

Just as the muscles attached to the dorsal surface control the action of the head, so
do the majority of those of the ventral surface give its various motions to the foot.
They are six in all. First, two central muscles, closely parallel (fig. 4, 5, 5), and each
divided into two branches at its upper end, run from nearly the middle of the lorica
down to the bottom of the foot, where they appear to be attached to one of the
toes. Next, two slighter ones (fig. 4, 6, 6), which pass up from the base of the toes,
one on each side of the foot, and then diverge right and left from the central pair to
points on the lorica about half-way between the median line and its edge. Besides
these, there is, attached to the upper end of the foot and on each side of it, a muscle
(fig. 4, 7, 7) which diverges still farther from the median line, and is fastened not far
from the margin. All these six can act together, so as to draw the foot suddenly
within the lorica ; while, by contracting some and relaxing others, the animal can
whisk the foot about, or, if the toes be adherent to any substance, can shake its whole
body vigorously from side to side—a favourite action.

The Nervous System and Organs of Sense.

The nervous system is represented by a bean-shaped cephalic ganglion (figs. 11, 1, *gn*)
seated within the head on its dorsal side. Its substance is marked with what appear to
be the hexagonal boundaries of cells. It is two-lobed posteriorly, and on the niche
between the lobes lies the crimson eye (fig. 11, *e*).

The pigment is distinctly curved round both sides of the niche so as to lie on each
lobe, and to give rise to the notion that the eye may really represent a pair fused into
one. I have not detected in this Rotiferon the clear spherical lens which is so plainly
visible in some of the others.[1] Attached to the upper portion of the cephalic ganglion
on the dorsal side is a conical and very flexible tube (fig. 1, *a*), whose broad base rests
on the ganglion, and whose free end passes through the sinus in the lorica between the
two longest spines. This is the dorsal *antenna*. A bulb armed with motionless setæ
completely closes the orifice of the tube ; and is so attached to its rim, that when this
bulb is withdrawn by the contraction of a muscular thread fastened to its base, the
tube is drawn down also by being infolded like the finger of a glove (fig. 15).

There are two other setigerous bodies, close to the dorsal surface of the lorica,
and with the setæ protruding through the surface. They are near the margin on each
side, a little above the line of greatest breadth (fig. 1, *a'*). They are rocket-shaped structures, the cylindrical heads carrying on their blunt, rounded, outer ends, radiating setæ ;
and giving off, from their inner and pointed ends, cords which can be traced but a little
way below the surface.

But the list of the tactile organs of *Brachionus* is not yet exhausted. From the two
spaces on the dorsal side of the corona lying between the three large setigerous prominences (fig. 1, *sp*) rise two papillæ, each bearing a long and very flexible style (fig. 1, *ts*).
A similar style (fig. 2, *t's'*) is placed on the ventral surface, just within the rim of each ciliary
circlet ; and another pair on the top of the central prominence. The whole six are
very mobile ; and, from the way in which they seem to be used to explore in all directions, there is little doubt that they are organs of touch.

---

[1] As in *Pedalion, Conochilus, &c.*

The Reproductive System.

The reproductive system of the female is only too conspicuous; as the presence of a large ovary (figs. 1, 2, *oy*), and of one or two opaque ova in different stages of growth, frequently obscures the sight of the other organs.

The ovary is studded with large and rather irregularly shaped germs (figs. 1, 2, *g*); and the ova (fig. 1, *om*), so long as they are within the body, are dark, granular, and homogeneous. The ovary opens doubtless by an oviduct into the cloaca, but this I have not been able to see. This is the arrangement that exists in other cases wherever I have seen the oviduct. The mature ovum is expelled very quickly; and the egg often remains attached to the animal by a thread till the young escapes by rupturing the shell (if it may be so termed) in which it is inclosed. If, however, a *Brachionus*, with two or three eggs attached, be held in the compressorium, it will frequently free itself by pushing with its foot against the eggs, and so breaking the threads.

The Egg.

Of the eggs and their development it is unnecessary here to say more than that *B. rubens* has two kinds of female eggs, nearly of the same size; of which the one has a smooth, transparent, membranous covering or shell, while the other has a thick opaque shell, ornamented with hemispherical knobs. The former is the ordinary " summer " egg, and the latter the so-called " lasting," " winter," or " ephippial " [1] egg. In shape, too, the ephippial egg differs from the ordinary female one. It is much bigger at one end than the other, and at the smaller end there is a projecting neck and cover (fig. 16).

The male eggs are smooth and semi-transparent like the ordinary female egg; but are somewhat rounder in shape, and barely half the size. They occur, too, in larger clusters; for while it is usual to see only two or three smooth female eggs, or one ephippial egg, attached to the mother, no fewer than eight or ten male eggs may be seen carried about together.

The Male.

The young female *Brachionus* when hatched resembles its mother: but the young male is a widely different animal (figs. 13, 14). It is about one-third of the length of the adult female's lorica, and it has neither lorica, mastax, jaws, nor stomach.

The head bears a simple circle of long cilia, and there is a red eye on a cephalic ganglion placed just as it is in the female. The vascular system, with its tortuous canals, vibratile tags, and contractile vesicle, is also present; and the foot is furnished with the usual pair of glands; but of a nutritive system there is not the slightest vestige.

Nearly half the body is occupied by a great *sperm-sack* (figs. 13, 14, *ss*), in which under favourable circumstances the spermatozoa themselves may be seen in motion. The sperm-sack ends in a short protrusile tube, the *penis* (figs. 13, 14, *p*), ciliated at the end, and placed just above the foot.

In the larger and more transparent males of other genera—notably in that of *Asplanchna Ebbesbornii*—there are special muscles for drawing back the protruded penis; and, though I have failed to see these in the male of *B. rubens*, it probably possesses a similar structure.

This strangely unfurnished creature leads a brief life of restless energy, now darting from place to place, so swiftly that the eye can scarcely follow it, and now whirling round as if anchored by its curved foot and penis.

It often circles round the female, attaching itself now here, now there, and forcing its companion to waltz round and round with it, from the top of the phial to the bottom. With animals so active and so small it is difficult to be certain of having seen actual

---

[1] So named by Prof. Huxley from the resemblance to the lasting-eggs of the *Daphnia*. These latter were called " ephippial " by Müller from their shape, which is that of an *ephippium*, or saddle.

coitus, but Mr. Gosse had this good fortune in the case of a closely allied species, *B. pala.* He says :

" I collected about a dozen females, half-grown and adult, and placed with them two lively males that had been hatched during the night. I directed my attention principally to one of these, as I could not watch them both. It soon came near one of the females, when it seemed to become animated by a sort of frenzy ; describing with excessive rapidity a circle, of which its head formed the circumference and its foot the centre. The extremities were incurved in the direction of its circular movement. After a while it left off, and began to play about the body of a female, moving over and round the lorica, while she whisked about the foot, as if to lay hold of him ; at length she drew in her foot, and that of the male appeared to adhere to it ; and I distinctly saw the thick penis presented to the cloaca, and for a moment inserted about half its own length ; then it was instantly drawn out, and the male began his frenzied gyrations again." [1]

It is obvious, even from the brief account here given, that several highly interesting questions arise concerning the reproductive system of the Rotifera. For instance, in what respects, besides outward form and covering, do the ephippial eggs differ from the smooth female eggs ? and what leads to their production ?

What part, if any, does the male play in these differences ? and how is it that one female produces none but male, while another has none but female eggs ?

Again, why are the males of so strange a structure ? and why do they appear only for a short time during the year ? and is their appearance due to external causes, or are they the inevitable completion of a cycle of reproductive changes ?

Unfortunately it is much easier to ask these questions than to answer them. The observations that have been made on these points are but scanty ; and, to some extent, contradictory ; and the difficulties in the way of persistent investigation are by no means slight.

[1] " On the Diœcious Character of the Rotifera." *Phil. Trans.* 1857, ¶ 19.

CHAPTER II.

THE HISTORY OF THE LITERATURE CONCERNING THE ROTIFERA.

Multum egerunt, qui ante nos fuerunt, sed non peregerunt. Multum adhuc restat operis, multumque restabit; nec ulli, nato post mille secula, præcludetur occasio aliquid adhuc adjiciendi.—SENECA, *Epist.* lxiv.

Cæterum nullius in verba jurans, aliorum inventa consarcinare haud institui; quæ ipse quæsivi, reperi, debitaque attentione et patientia repetitis vicibus, diversoque tempore annorum serie observavi, propono.—O. F. MÜLLER, *Verm. Terrest. et Fluv.* præf.

CHAPTER II.

THE HISTORY OF THE LITERATURE CONCERNING THE ROTIFERA.

IT is nearly 200 years since Mr. **JOHN HARRIS**, F.R.S., Rector of Winchelsea, published in the "Philosophical Transactions"[1] the following "Microscopical Observations" on a drop of some rain water which "had stood in a gallipot in his window" for about two months:

"I saw here an animal like a large maggot, which could contract itself up into a spherical figure, and then stretch itself out again; the end of its tail appeared with a forceps like that of an earwig; I could plainly see it open and shut its mouth, from whence air-bubbles would be frequently discharged. Of them I could number about four or five, and they seemed to be busy with their mouths as if in feeding."

This description is but vague; and yet it is very probable that the animal which the rector found in his gallipot was a *Rotifer*: and if so, this is the earliest notice that we have of the class.

A few years later, in 1703, **LEUWENHOEK** published, in the same "Transactions,"[2] an account of some animalcules, living in sheaths, that he had found at Delft, on green weed brought by the flooding of the Maes from Delft Haven. These little creatures were almost certainly *Limnias ceratophylli*. Leuwenhoek gives figures of one, and notices its bearing "two wheels thick set with teeth as the wheel of a watch." In a later paper[3] in the "Philosophical Transactions" he gives a much better account of *Limnias*, with greatly improved figures. He states that when the two wheels are viewed frontally they are seen to be continuous and to form but one; and he adds a figure (Pl. B, 4) of the corona so seen, and notices that the ciliary waves run right round the whole wreath in the same direction.

Leuwenhoek next describes an animal "that has a receptacle or little house composed of round bubbles," and furnished with "surprising wheelwork" of four parts, three of which only were shown in his figure, "the fourth being almost hid from sight." One glance at the figure of the animal (Pl. B, 1), and at its corona (Pl. B, 2), shows us that he has had *Melicerta ringens* under his microscope.

His next paper[4] on the same Rotiferon is pleasant reading; for it vividly recalls the shock of delighted surprise with which every naturalist first enters into the enchanted world beneath the waters.

"I took notice," he says, "of the surprising figure of an animalculum, fixed in a little scabbard or sheath, fastened to some of the small green weeds found in ditches of water. And, as often as I viewed these animalcula and showed them to others, we could not satisfy ourselves with looking on such surprising objects; and the more because we could not conceive how so strange a motion, as they all had, could be performed; as also what should be the use of such a motion." Leuwenhoek also relates how *Melicerta* makes its tube pellet by pellet; and this is his account of it. "I viewed one of these animalcula a good while together; and observed, several times, one after another, that when the animalculum thrusts its body out of the sheath or case, and that the

---

[1] *Phil. Trans.* vol. xix. 1696.   [2] *Phil. Trans.* No. 283, 1703.
[3] *Phil. Trans.* No. 295, 1705.   [4] *Phil. Trans.* No. 337, vol. xxviii. 1713.

wheel-like or indented particles moved in a circle, at the same time out of a clear and transparent place a little round particle appeared, which, without nicely viewing, could hardly be perceived; which particle growing larger, moved with great swiftness as it were about its own axis, and continued without any alteration in its place, till the animalculum had drawn part of its body back into its sheath; in doing which it placed the said round particle on the edge of its sheath, which thus became augmented with a round globule; and whereas the animalculum had placed the said globule on the east part of its sheath, another time it fixed it on the south or north side; by which means the sheath was regularly increased on all sides."

Of course this is but a rough sketch of the machinery and actions of the famous tube-maker; but it is a vigorous one, and true to the life as far as it goes.

With equal truth and vigour does Leuwenhoek describe, in the same paper, the transformations of *Philodina roseola*;—that creature, whose powers of lying dormant for rotiferous ages, and of then coming to life again, have made it as famous as the "Seven Sleepers."

To anyone with a sense of humour it must be delightful to read the following paragraphs of Leuwenhoek's paper; and then to reflect how his discoveries have been repeatedly re-discovered; and how again and again they have been challenged, confirmed, forgotten, and once more discovered. In fact, the Philodine has been the cause of a dispute which has all the marvellous properties of the Rotiferon itself. For it periodically goes to sleep and revives again, just as *P. roseola* does; but with a difference. The Rotiferon, when it awakes after its long sleep, takes up its life at the point where it left it off; and ultimately "gives up its murmuring breath" after an existence of three or four years; but the dispute concerning it invariably begins again both *de novo* and *de ovo*; and having already lasted, with periodical intermissions, for nearly 200 years, evidently bids fair to last for 200 years more; as it has only a short time ago awakened once more, as fresh and as vigorous as ever, and found its way into our daily papers.

" I discovered," says Leuwenhoek, "several animalcula that protruded two wheels out of the fore-part of their body as they swim, or go on the sides of the glass. . . . This sort I found, in great numbers, in the gutter water which had stagnated some days in the small pits or cavities of the lead. . . . In October 1702, I caused the dirt of the gutters, when quite dry, to be gathered together, and taking a small quantity of it, I put it into a paper on my desk; since which time I have often taken a little of it, and poured on it boiled water after it had stood still till it was cold, that I might obviate any objection that should be made, as if there were living creatures in that water. These animalcula, when the water runs off or dries away, contract their bodies into a globular or oval figure. . . . In the month of September I put a great many of the last-mentioned animalcula into a wide glass tube, which presently placed themselves on the sides of the glass; whereupon, pouring off the water, I then observed that several of the animalcula, to the number of eighteen or nineteen, lay by each other in a space of coarse sand; all of which, when there remained no more water, closed themselves up in a globular figure. Some of these animalcula were so strongly dried up that one could see the wrinkles in them, and they were of a reddish colour; a few others were so transparent, as if they had been little glass balls, that, if you held them up between your eye and the light, you might move your fingers behind them, and see the motion through their bodies. After these animalcula had lain thus dried up a day or two, in an oval or globular form, I poured some water into the glass tube; whereupon they presently sank to the bottom, and after the space of about half an hour began to open and extend their bodies, and, getting clear of the glass, to swim about the water. . . . In the month of October, before the dirt of the leaden gutter was quite dried up, I took a handful of it, and laid it on a glazed earthen dish in order to preserve it. . . . Upwards of twenty-one months after, I took some of this dry stuff and infused it, both in cold water that had been boiled, and in rain water newly fallen; whereupon the animals began to show themselves in great numbers."

The only points of this much vexed question that Leuwenhoek passes over are:

(i) how the Rotiferon, when drawn up into a ball, resists the persistent baking of a summer's sun on the housetop, or the long drought of twenty-one months on the naturalist's desk, without parting with its own internal moisture ; and (ii) why only eighteen or nineteen of his Rotifera (those near the coarse sand) succeeded in rolling themselves up and surviving, while the rest perished.[1]

The investigations thus worthily begun by Leuwenhoek were carried on with much spirit by many other observers ; and, during the hundred and thirty-five years that elapsed before the publication of Ehrenberg's famous work, " Die Infusionsthierchen," no fewer than sixty of Ehrenberg's species, contained in thirty of his genera, were entered on the list of known Rotifera.

I have arranged the more striking forms of these in the following table, which classifies them under the heads of some of the families into which I have divided the Rotifera ;[2] and I have added the names and dates of their discoverers ; it will be thus seen how wide a ground had been covered by the early naturalists, since more than half the families have representatives in the table.

A List of some of the Rotifera discovered before 1838.

FLOSCULARIADÆ .	{ Stephanoceros Eichhornii	. .	. Eichhorn, 1761
	{ Floscularia ornata	. .	. Eichhorn, 1767
MELICERTADÆ	{ Melicerta ringens .	. .	. Leuwenhoek, 1703
	{ Limnias ceratophylli .	. .	. Leuwenbock, 1703
	{ Lacinularia socialis .	.	. Brady, 1755
	{ Megalotrocha albotlavicans .	.	. Rösel, 1755
PHILODINADÆ	{ Rotifer macrurus .	. .	. Baker, 1755
	{ Philodina roseola .	. .	. Leuwenhoek, 1702
HYDATINADÆ	. Hydatina senta .	. .	. Müller, 1773
SYNCHÆTADÆ	. Synchæta Baltica .	. .	. Baster, 1759
NOTOMMATADÆ	. Notommata tigris .	. .	. Müller, 1786
TRIARTHRADÆ	. Triarthra longiseta	. .	. Eichhorn, 1775
	{ Brachionus pala .	. .	. Joblot, 1718
BRACHIONIDÆ	. { Brachionus urceolaris .	. .	. Hill, 1751
	{ Anuræa striata	. .	. Müller, 1776
PTERODINADÆ	. Pterodina patina .	. .	. Eichhorn, 1775
EUCHLANIDÆ	. Euchlanis dilatata	. .	. Eichhorn, 1775
RATTULIDÆ .	. Mastigocerca carinata .	.	. Müller, 1786
DINOCHARIDÆ	{ Dinocharis pocillum	. .	. Müller, 1776
	{ Stephanops lamellaris .	.	. Müller, 1786

It will be seen that the names of Eichhorn and Müller occur much more frequently than that of any other observer in this list, and their works on the Rotifera deserve, I think, a special notice.

JOHANN CONRAD EICHHORN was the pastor of St. Catharine's church at Danzig, and his book on the " Natural History of the Smallest Aquatic Animals "[3] was published in 1781. Though small, it is a most interesting work. He gives figures and descriptions of about a dozen Rotifera that can be identified, including *Stephanoceros*, *Floscularia*, *Melicerta*, *Triarthra*, *Dinocharis*, *Actinurus*, *Euchlanis*, and *Pterodina* ; and of most of his species he was the discoverer.

His enthusiasm is delightful. " I have devoted myself," he says, " to this invisible world, which yields itself to our ken only under the magnifying glass ; and I have, for eleven years, spent my leisure moments on it, so far as my professional duties would permit, in order to know God in His smallest and invisible works ; and I have found Him very great therein. Not the great works only, those vast heavenly bodies —

[1] The question will be found fully discussed farther on in the general account of the family Philodinadæ.

[2] See chap. III. On the Classification of the Rotifera.

[3] *Beiträge zur Naturgeschichte der kleinsten Wasserthiere.*

not those huge animals, on earth as well as in sea, who can scarcely drag the
weight of their frames—not these alone declare the glory of the Almighty. No ! the
smallest also show, just as distinctly, the perfections of their Creator. Yea ! one may
say, these even more than those ! A great church clock is certainly a wonderful machine,
but a pocket watch—a watch in a ring—is yet more so, and conduces to the greater fame
and glory of its maker."

His description of his chief discovery, that of his Crown Polyp (*Stephanoceros
Eichhornii*) is very amusing. " I found," he says, " this extraordinary and marvellously
formed animal first in 1761, on July 20, on a water plant, which had been standing some
weeks in water. I saw that there was something on the plant which was quite unknown
to me. I moved the glass, in order to see if it was something alive, and if it would draw
itself together, which happened, to my delight ; therefore I examined it through a lens,
but it appeared to me, through this, just like an orange flower which was not yet closed,
but which now drew itself together, and now outspread itself. All this stirred up in me
a great desire to see this new animal under the glass, but that required skill to get it
out, as the glass vessel in which it was, was nearly an ell high, and this animal was right
at the bottom. I tried first with the quill of a feather to bring it to the top, but it was
continually lost to the eye by shutting itself up. At last I succeeded with a little wire
hook in drawing out the plant on which it obviously was, and as soon as I could reach
it with the scissors I snipped off a tiny stem, and that brought me out the whole animal
unharmed. I placed it at once under the magnifying glass, and saw this matchless
creature as it is shown in the engraving." [1] What a pleasant picture this is of the
grave pastor fishing away with a quill pen to fetch up *Stephanoceros* from the bottom of
a glass beaker a yard and a quarter high !

About the same time as Eichhorn, flourished the great Danish naturalist, **OTHO
FREDERIC MÜLLER.** He was an excellent botanist and zoologist, and published
works on many subjects. He wrote on the Flora and Fauna of Denmark, on Fungi, on the
Hydrachnæ, and on Fresh-water and Marine Worms ; but his chief delight was in the
Infusoria, and his posthumous work, " Animalcula Infusoria Fluviatilia et Marina, &c."
1786, was the first that brought this new kingdom to the knowledge of the naturalist.

The " Animalcula Infusoria " contains the descriptions and figures of about fifty
Rotifera, among which are *Lacinularia, Hydatina, Scaridium, Triarthra, Brachionus,
Anuræa, Pterodina, Euchlanis, Dinocharis, Stephanops,* and *Mastigocerca*. More than
half of Müller's species were new when published ; and his figures, taken from life, are
beautifully drawn on copper by himself. Of course there is a great lack of detail in the
drawings of the internal structure of the animals, but they are an immense advance on
those of Eichhorn, the outlines being usually both spirited and faithful.

Müller's text, too, is as good as his figures. It is the work not only of a naturalist,
but of a thoughtful and learned man ; and both the " Animalcula Infusoria " and his pre-
vious work, " Vermium Terrestrium et Fluviatilium," abound with admirable and striking
passages. In the latter, he thus begins his dissertation on the Infusoria : " The world
of the invisible, a world shut to our ancestors, was first entered about a hundred years
ago. It breeds monsters of unheard-of form and manner of life, it abounds in miracles
as much as do the remote Indies ; but is explored with lesser peril, for it lies everywhere
at our very feet, and is not sought out for gold.

" Each was explored with great slaughter of its inhabitants ; the one often resisted
by wasting the lives of its aggressors, the other had no defence but patience.

" *This* we owe to the needle, which joined two hemispheres together ; *that* to the
lens, which images alike the solar spots and the infusoria, the widest apart of all things.

" In this interval what indeed is great, what little ? Man : for he thinks and suffers."

L. JOBLOT styles himself, " Professeur Royal en Mathématiques, de l'Académie Royale
de Peinture et Sculpture, demeurant sur le Quay de l'Horloge du Palais, au gros Raisin."

---

[1] See Pl. B, fig. 11.

His book, published in 1718, consists of two parts. In the first he describes various forms of microscope, and the best way of using them; in the second he details, from his diary, the results of a series of experiments made with infusions of various plants and substances. The list of his infusions is long and curious. He tried pepper, roses, rhubarb and senna, maize, violets, mushrooms, hay, raspberry stalks, celery, knapweed, fennel, straw, marigolds, melons, tea, oak bark, &c. &c., and even found an animalcule that,—

for saving charges,
A peeled sliced onion eats, and tipples verjuice.

He scoffs at the notion that living animals could be produced by the putrefaction of anything, and is confident that the infusion of each substance produces its own peculiar animals. He supposes that eggs are laid on these substances "by a countless number of very little animals that fly or swim in the air close to the ground," and who "let fall their eggs and little ones as they course backwards and forwards in the air," particularly in the spots where they were stopped by the vapours escaping from their favourite plants.

He was the first to discover the genus *Brachionus* (*B. pala* and *B. amphiceros*), and he found in his infusions various species of the genus *Rotifer* (probably *R. vulgaris* and *R. citrinus*) as well as, possibly, a *Lepadella* and a *Monostyla*.

His figures are grotesque enough; and he damaged his reputation as a sober naturalist by sketching a six-legged creature with "tout le dessus de son corps couvert d'un beau masque bien formé, de figure humaine, parfaitement bien fait." A fierce moustachioed face it is, and, as Joblot adds, "couronné d'une coëffure singulière." His names for the animals are as odd as his figures: he has "top-knots," "bagpipes," "dandies," "tortoises," "kidneys," and even "crowned and bearded pomegranates"—the last strange title being given to his new discoveries, the *Brachioni*.

In spite of all these absurdities his written descriptions are often vivid and accurate, and he is a shrewd observer. For instance, he notices how cleverly the Rotifera swim without jostling each other; and he concludes that "though we cannot see them, they must have eyes, and those very good ones." He describes the restless movement of *Brachionus* swaying from side to side as it thrusts about its long foot, and observes that "there are some females who carry only one egg, some that carry two, and some as many as six, which, however, is not common; and when there are so many eggs they are smaller in size than they are when they are fewer."

It is thus clear, both from his description and his figure, that this first discoverer of *Brachionus pala* had already lighted on a female carrying a cluster of male eggs. Again, he discerns the difference, in size and colour, between *Rotifer vulgaris* and *R. citrinus*; and describes their leech-like movements, their telescopic joints, and their constantly moving jaws. I need hardly add that he considers the latter to be the heart.

His comments on his discoveries are as characteristic as his figures and theories. We have seen the Danish naturalist exulting in the human intellect which, armed with one simple weapon, attacks alike the distant planets and the invisible infusoria; and finding even greater reason for his pride in the very weakness and ills that flesh is heir to. The Prussian pastor, too, is as enthusiastic after eleven years' study as he was when he first began; and, as if he would add another verse to the Benedicite, cries to all the creatures of the invisible world, "Bless ye the Lord; praise Him and magnify Him for ever." And Professor Joblot is also enthusiastic, but his strain is pitched in a different key. He says of one of his infusions that "it gives rise to a most delightful spectacle, so curious to see and watch that I do not think that the diversion of the play, of the opera with all its magnificence, of rope-dancing, tumbling, or of the animal-combats which we see in this superb city, ought to be preferred before them."

It would be unfair to M. Joblot not to add that he is capable of better things, as the following extract shows:

"There is nothing despicable in nature; and all the works of God are worthy of our respect and admiration, especially if we take heed to the simplicity of the means by which God has made and preserved them. The smallest gnats are as perfect as the hugest animals, the proportions of their limbs are equally just, and it seems as if God had even wished to give them more ornaments than He has to the greater creatures, in order to make up to them for the smallness of their bodies. They have crowns, tufts, and other adornments on their heads, which surpass all that female luxury has invented ; and we may say that those who have looked only with unaided eyes have seen nothing so beautiful, so fitting, nor even so magnificent, in the palaces of the greatest princes, as that which the microscope shows on the head and body of a simple fly."

About forty years after Joblot, **HENRY BAKER**, F.R.S., published a somewhat similar work. The first volume, " The Microscope made Easy," treats of the instrument itself ; while the second volume, " Employment for the Microscope," describes the various things that may be seen with it.

In the second volume he gives an elaborate account, with figures, of what I believe to be *Philodina roseola* ; as well as descriptions and drawings of *Rotifer macrurus, Brachionus pala, B. urceolaris, B. Bakeri,* and probably also of *Euchlanis triquetra ;* and of these six species the second and last had not been described before.

His drawings are vastly superior to those of Joblot, especially his figures of the *Brachioni.* He notices and introduces into his figure the long vibrating styles which crown the head of *B. pala,* as well as its winter eggs. He failed, indeed, to understand the lorica of *Euchlanis ;* but that is no wonder, for he has had many to bear him company.

It is unnecessary for me to say more of a book that is still within everyone's reach ; but there is one admirable passage in his preface that I must give myself the pleasure of quoting.

" That man is certainly the happiest who is able to find out the greatest number of reasonable and useful amusements, easily attainable and within his power ; and, if so, he that is delighted with the works of nature, and makes them his study, must undoubtedly be happy ; since every animal, flower, fruit, or insect, nay, almost every particle of matter, affords him an entertainment. Such a man never can feel his time hang heavy on his hands, or be weary of himself, for want of knowing how to employ his thoughts ; each garden or field is to him a cabinet of curiosities, every one of which he longs to examine fully ; and he considers the whole universe as a magazine of wonders, which infinite ages are scarce sufficient to contemplate and admire enough."

In Plate B, I have given copies of some of the old figures drawn by these authors, and if the reader will compare them with **EHRENBERG**'s drawings of the same animals, he will see at a glance why the Prussian naturalist's work [1] swallowed up as it were the very memory of all his predecessors. Instead of feeble, inaccurate drawings, in which the internal structure was represented by mere blots and patches, Ehrenberg gave excellently drawn figures full of accurate details ; and at the same time described the animals themselves with wonderful exactness, considering the very great number that he studied unaided.

Nor was this all : he had such a grasp of the whole subject, such a minute personal knowledge of the living animals themselves, that he invented a system of classification which has held its own for nearly fifty years.

In addition to its other merits, Ehrenberg's splendid work added more than a hundred new species to those already known, containing among them such remarkable forms as *Conochilus volvox, Notommata clavulata, N. copeus, N. centrura, Diglena grandis, Polyarthra platyptera, Noteus quadricornis, Microcodon clarus, Œcistes crystallinus,* &c.

Three years after the publication of " Die Infusionsthierchen," **DUJARDIN** published his " Infusoires " as one of the volumes of the Histoire Naturelle des Zoophytes in the " Suites à Buffon." The last part of this volume, being one-sixth of the whole, is devoted to the " Systolides " or Rotifera. His book is mainly critical, and, so far as I can find, contains little on the Rotifera that was new, except his observations on *Albertia* and *Lindia,*

[1] *Die Infusionsthierchen ;* Leipzig, 1838.

His criticisms are shrewd and often just; he points out that Ehrenberg's respiratory tube is probably an antenna; suggests that the convoluted tubes, flickering tags, and contractile vesicle of the vascular system have a respiratory function; calls attention to the varying forms of the mastax and trophi as good characters for classification; and conjectures that the perivisceral fluid is the true analogue of the blood.

On the other hand, he could not see *Floscularia's* tube; could not make out the striated muscles in any Rotiferon, even in *Pterodina patina*, of which he gives a figure; could see indeed no difference between the muscles and the nerves; doubted the existence, as specialised structure, of either the one or the other; and from want of personal acquaintance with them, affirmed the identity of many of Ehrenberg's species, which are undoubtedly distinct.

But although he has small claim to be considered either an original or an accurate observer of the Rotifera, he made one happy hit in his attempted classification, which will be detailed elsewhere.

Since Dujardin's time the more noteworthy essays that have been published on various portions of our present subject are by Mr. P. H. Gosse, F.R.S.; Dr. F. Leydig; Professor T. H. Huxley; Herr C. Vogt; Dr. F. Cohn; Dr. W. Moxon, F.L.S.; Dr. W. Salensky; Dr. S. Bartsch; and Herr Karl Eckstein.

Mr. **GOSSE**, in his paper, " On the Anatomy of *Notommata aurita*," [1] described with minuteness the organization of this common species, so that the essay became, as it were, a key to the structure of the majority of free-swimming Rotifers. His next treatise, " On the Structure, Functions, and Homologies of the Manducatory Organs in the Class Rotifera," [2] is illustrated with a great many drawings of the mastax and trophi of various species; and discusses the changes that they undergo, in passing from the typical to the most aberrant forms. It is in this treatise that Mr. Gosse contends that the dental organs of the Rotifera are true mandibulæ and maxillæ; and that the mastax is a mouth; and assigns to the Class a position among the *Articulata*. In a subsequent paper, " On the Diœcious Character of the Rotifera," [3] Mr. Gosse extended this character from a single genus, *Asplanchna*, to five others; and trebled the number of the known diœcious species. Some years later, Mr. Gosse began, in " Contributions to the History of the Rotifera," [4] a general account of the whole class, arranged according to a classification of his own, and continued it so far as the *Flosculariadæ*, *Melicertadæ*, and *Notommatina*, illustrating each family with descriptions and figures of certain selected species. This work, however, owing to the cessation of the periodical, was never completed.

Dr. **F. LEYDIG**, in " Ueber den Bau und die systematische Stellung der Räderthiere," [5] after a full description, accompanied with figures, of many species, three of which are new, proceeds to deal with the structure of the Rotifera as a class, and to arrange them in a system of his own. He further discusses their true position in the animal kingdom, and assigns them a place among the *Crustacea*.

Professor **HUXLEY**, in his paper, " On *Lacinularia socialis*; a Contribution to the Anatomy and Physiology of the Rotifera," [6] takes this Rotiferon as his text, and, while minutely describing its structure, discusses various questions concerning that of the whole class. He enters into the general relations of the Rotifera to other animals, and arrives at the conclusion that they are permanent forms of Echinoderm larvæ.

Herr **C. VOGT**, in his treatise, " Einige Worte über die systematische Stellung der Räderthierchen," [7] combats Leydig's reasonings and conclusion on the position of the Rotifera, and affirms that they must be classed among the *Vermes*.

In Dr. **F. COHN**'s essay, " Ueber die Fortpflanzung der Räderthiere," [8] the males and females of three species are minutely described, especially with reference to their reproductive organs; and the general question of the reproductive system of the whole class is also discussed.

Dr. **W. MOXON**'s ' Notes on some Points in the Anatomy of the Rotatoria' [9] call

[1] *Trans. Micr. Soc.* 1852.   [2] *Phil. Trans.* 1856.   [3] *Phil. Trans.* 1857.
[4] *Popular Sci. Rev.* 1862 and 1863.   [5] Leipzig, 1854.   [6] *Trans. Micr. Soc.* 1853.
[7] *Sieb. u. Köll. Zeits.* 1855.   [8] *Sieb. u. Köll. Zeits.* 1856.   [9] *Trans. Linn. Soc.* 1861.

attention to the right use of the terms " dorsal " and " ventral " as applied to the Rotifera ; to the existence, in many species, of three antennæ, holding definite positions with respect to the dorsal and ventral surfaces ; to the true nature of the entrance to the crop of *Floscularia* ; and to the structure and function of the vibratile tags.

Dr. **W. SALENSKY's** paper, " Beiträge zur Entwicklungsgeschichte der *Brachionus urceolaris*," [1] traces the gradual changes in the ovum, from its first division into two unequal spheres, through its complete segmentation, to the formation of the germinal layers, and the evolution of the various organs of the completed embryo. Although the paper deals with only a single species, it is our principal contribution to the Embryology of the Rotifera.

Dr. **S. BARTSCH**, in " Rotatoria Hungariæ," 1877, and Herr **KARL ECKSTEIN**, in " Die Rotatorien der Umgegend von Giessen," [2] have published treatises containing descriptions and figures of local Rotifera (forty in Hungary and fifty at Giessen), including two or three new species ; as well as new classifications of the whole class ROTIFERA.

Of Dr. **BARTSCH's** work I can say but little, as it is written unfortunately in Hungarian. His figures, though somewhat archaic, are well worth the studying ; and he gives drawings and descriptions (happily this time in Latin) of six new species, of which two had been recorded in England some years before.

Herr **ECKSTEIN** also gives many interesting details of his local species, of which two are new ; and adds a general discussion of the structure, development, affinities, and classification of the whole class. His treatise also contains useful lists of synonyms, as well as a good bibliography of the subject.

The last edition (1861) of **PRITCHARD's** " History of Infusoria," by Dr. Arlidge and others, is a work differing in character from any of the above. About one-sixth of it is devoted to the Rotifera, and contains descriptions of the whole of the then known species, illustrated by a great many figures. Both the descriptions and the figures have been mainly taken from Ehrenberg's work, which is closely followed throughout ; but they have been supplemented by others taken from the various treatises mentioned above.

As a compilation, it is not only the best, but almost the only, English work on the subject. It contains, moreover, an admirable and exhaustive treatise on " The General History of the Rotifera " as a class, dealing minutely with their structure, reproduction, development, systematic position, and classification.

This original and most valuable essay may be said to be indispensable to all students of the Rotifera, bringing together, as it does, into one point of view, the opinions of all the best observers, on the many vexed questions that these little creatures have given rise to, not only as to their organization and development, but as to their relations to the rest of the animal kingdom.

A full list of works on the Rotifera, including numerous papers that have been published in various scientific periodicals, will be found at the end of this work.

---

[1] *Sieb. u. Köll. Zeits.* 1872.　　　　　　[2] *Sieb. u. Köll. Zeits.* 1883.

CHAPTER III.

ON THE CLASSIFICATION OF THE ROTIFERA.

Omnis enim systematica divisio claudicat lacunisque laborat; optima est, quæ pauciorihus horret, documentum satis splendidum, mortales non e vero visionis puncto Naturam contemplari.—O. F. MÜLLER.

Tous les ordres des êtres naturels ne forment qu'une seule chaîne, dans laquelle les différentes classes, comme autant d'anneaux, tiennent si étroitement les unes aux autres, qu'il est impossible aux sens, et à l'imagination même, de fixer précisément le point, où quelqu'une commence ou finit. — LEIBNITZ.

Dum inter ea, quæ determinatis characteribus discreta, et certis quasi limitibus inclusa sunt, semper intermediæ quædam species reperiuntur, quæ, utrinsque proxime accedentis speciei, aliquid possideant, et ita copulationem quasi duarum diversarum specierum constituant; colorum ad instar, qui ita commiscentur et quasi confluunt, ut nemo veros cujusque fines determinare possit.—J. BASTER.

CHAPTER III.

ON THE CLASSIFICATION OF THE ROTIFERA.

FOUR attempts have been made to improve upon Ehrenberg's classification : viz. that of Dujardin in 1841, of Leydig in 1854, of Dr. S. Bartsch in 1877, and of Herr K. Eckstein in 1883. I do not intend to discuss here the various merits and faults of these five systems; it is enough to say that they all seem to have the fault of needlessly bringing together animals that are different in structure, while separating others that closely resemble each other ; I say "needlessly" because perfect classification appears to be an impossibility, except at that fleeting stage of our knowledge when none but the commoner genera are known to us. These usually differ from each other in a marked fashion, the very fact of their wide-spread co-existence being perhaps due to their differing so as not to interfere with each other.

When, however, continued search has brought to light the rarer forms, these usually prove to be links between the more common ones ; and then the troubles of the classifier begin. For these strange forms, which are the delight of the naturalist, are the classifier's despair. Do what he will, no system that he can devise will put into Nature those sharp divisions and well-marked gaps that are so dear to him, but of which she knows nothing.

Nature has but one law, that of infinite variety ; and the utmost that the classifier can do is to group his animals as well as he can round certain typical forms, content to have the symmetry of his plans and the sharpness of his definitions marred by forms that perversely bear the characteristics of two or three of his types, in nearly equal proportions.

He may take comfort, nevertheless ; for, even if he had been able to invent a thoroughly satisfactory classification, it is from the nature of the case written in sand. He can never say as he throws down his pen :

Exegi monumentum ære perennius ;

for it is almost certain that the fresh discoveries of the next ten years will require his work to be re-cast ; and no higher praise could be given to Ehrenberg's system than that, in spite of new discoveries and its own obvious faults, it has reigned alone for nearly five times the usual period.

The Rotifera may first be divided into four natural orders, according to their modes of locomotion, and the structure of the foot. The first of these ideas appears in Dujardin's classification, and the second in Leydig's, and they are both excellent ; for there are Rotifera that swim by means of their ciliary wreath, and skip by the help of their arthropodous limbs; Rotifera that swim only with their wreath ; others that swim and creep like a leech ; and lastly, some that, when adult, are stationary. Moreover, in three of these four orders there is only one form of foot in each order, and that form is unlike those of the other two ; and although in one order there are more forms of the foot than one, still they are all unlike the forms of the other three.

Nor is this all. The natural character of these four orders is further shown by there being other important points of structure, in which the animals comprised in each order at the same time resemble each other and differ from those of the other orders.

I have named these orders as follows :

I. RHIZOTA (the rooted). Fixed when adult.
II. BDELLOIDA (the leech-like). That swim with their ciliary wreath, and creep like a leech.
III. PLOIMA (the sea-worthy). That only swim with their ciliary wreath.
IV. SCIRTOPODA (the skippers). That swim with their ciliary wreath, and skip with Arthropodous limbs.

Now, the creatures contained in these orders, with a few exceptions, differ from each other, first in their habits, and secondly in the following points :

I. In the structure of the foot.
II. In the arrangement of the ciliary wreath.
III. In the form of the trophi.

I. In the structure of the foot.

(1) The *Rhizotic* foot is transversely wrinkled, and ends in a sort of sucking disk (fig. 16) or cup (fig. 17). It is not retractile within the body, it never has telescopic joints, nor is it ever furcate.

Fig. 16. Rhizotic foot. (*Floscularia campanulata*)   Fig. 17. Rhizotic foot. (*Melicerta ringens*)   Fig. 18. Bdelloidic foot. (*Rotifer vulgaris*)   Fig. 19. Scirtopodic foot. (*Pedalion mirum*)

(2) The *Bdelloidic* foot (fig. 18), on the contrary, is telescopic, retractile, furcate, and is never transversely wrinkled, nor terminated by a sucking disk.

(3) The *Scirtopodic* foot (fig. 19) is unique ; it is divided into two unconnected, smooth, jointless styles, each ending in a ciliated expansion.

(4) The *Ploïmic* foot is various in shape, but is always unlike that of any other order ; for—

(a) if transversely wrinkled, it is yet retractile within the body, and almost invariably furcate ;
(b) if jointed and furcate, it is not also telescopic ;
(c) occasionally it is absent altogether.

II. In the arrangement of the ciliary wreath.

(1) The *Rhizotic* ciliary wreath is of two forms :

(a) The first encircles the body twice, by bending on itself ; thus inclosing the mouth, and having a dorsal gap between the points of flexure, figs. 20, 21.

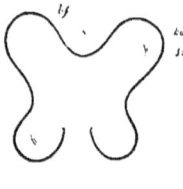
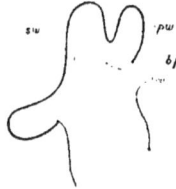
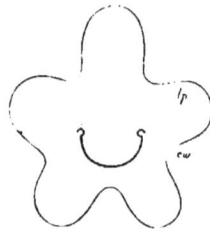

Fig. 20. Rhizotic wreath (a), front view. (*Melicerta ringens*)   Fig. 21.—Rhizotic wreath (a), side view. (*Melicerta ringens*)   Fig. 22. Rhizotic wreath (b), seen from above. (*Floscularia campanulata*)

[1] In figs. 20 to 25, *cw* is the ciliary wreath ; *pw* is the principal wreath ; *sw* is the secondary wreath ; *bf* the buccal funnel ; *lp* the lips.

(b) The second form is a simple segment of a circle, placed on the ventral side above the mouth, fig. 22.

(2) The *Bdelloidic* wreath is also of two forms :

   (c) The first, like the *Rhizotic* wreath (a), is a double wreath surrounding the body twice, and inclosing the mouth ; but, unlike the *Rhizotic*, it has two gaps instead of one, viz. a dorsal gap between the points of flexure, and a ventral gap in the upper wreath opposite to the mouth.

   (d) The second form of Bdelloidic wreath is a mere furring of the corona on its ventral surface, as shown in fig. 25.

(3) The *Ploïmic* wreath is very various in shape, but is never *Rhizotic*, while it is *Bdelloidic* only in one genus.[1]

Fig. 23.— Bdelloidic wreath (c), from above. (*Rotifer citrinus*)

Fig. 24.—Bdelloidic wreath (c), side view. (*Rotifer citrinus*)

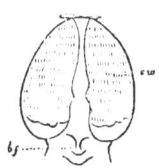

Fig. 25.—Bdelloidic wreath (d). (*Adineta rosa*)

(4) The *Scirtopodic* wreath is of *Bdelloidic* type.

III. In the form of the trophi.

If we disregard two genera [2] of the *Ploïma* (not one-fifteenth of the whole number of Ploïmic genera), we can then say that the first three orders differ also in the forms of their trophi. For—

(1) The *Rhizotic* trophi are either *malleo-ramate*,[3] fig. 26, or *uncinate*,[3] fig. 27.

(2) The *Bdelloidic* trophi are always *ramate*,[3] fig. 28.

(3) The *Ploïmic* trophi are of various forms ; but are never *Bdelloidic*, and are *Rhizotic* only in two genera.[2]

(4) The *Scirtopodic* trophi are of a Rhizotic type, being *malleo-ramate*, fig. 26.

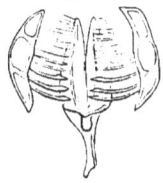

Fig. 26.—Malleo-ramate. (*Melicerta ringens*)

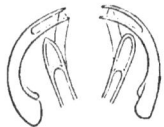

Fig. 27.—Uncinate. (*Stephanoceros Eichhornii*)

Fig. 28.—Ramate. (*Rotifer citrinus*)

Now, in reviewing the points of agreement and of difference in the four orders, we may at once set aside the fourth order, the *Scirtopoda*, as unmistakably separated from the others.

This order contains but one family, which has only one genus, and that genus itself consists of only one species.[4] In fact, it has been formed to contain that remarkable creature *Pedalion mirum*, which I discovered at Clifton in 1871. This

---

[1] *Pterodina.*   [2] *Triarthra* and *Pterodina.*

[3] For the explanations of these technical terms, see p. 29.

[4] I pass over for the present Dr. Schmarda's *Hexarthra polyptera*, and will discuss it under the *Pedalionidæ.*

Rotiferon has six hollow limbs continuous, in true Arthropodous fashion, with the body-cavity, and worked by opposing muscles passing down them, and is thus plainly linked to the Crustacea and Insecta. *Pedalion*, in fact, is a Nauplius larva, and is yet a Rotiferon.

Order IV., then, is sufficiently separated from the rest by its Arthropodous limbs, and by the use made of them; and of the other three orders, it has been shown above that, if we disregard some points of only two genera, we may say of orders I. II. III. that they differ *inter se* in their habits, and in the structure of their feet, trophi, and ciliary wreaths.

This seems a satisfactory first step towards classification; but it is only fair to the reader to warn him that it has been gained by omitting some parasitic Rotifera, as well as a few very troublesome forms, such as *Trochosphæra*, *Apsilus*, *Microcodon*, &c.

I have dwelt on the differences in structure, as well as in habits, between the four orders, in order to show that these four groups are natural; but I do not propose to use as ordinal characteristics any others than the mode of locomotion and the structure of the foot; and for this reason, that each of the first three orders has more than one form of the trophi, or of the ciliary wreath, or of both.

The *Rhizota*, for instance, have two forms of the trophi, and two of the ciliary wreath. The *Bdelloida* have two forms of wreath, and the *Ploima* have many different forms of both wreath and trophi.

Before I proceed to divide these four orders into families, I must, however, digress a little to explain and name the various types of trophi, as the classification partly depends upon them.

Mr. Gosse's treatise on " The Manducatory Organs in the Class Rotifera " essays to show that these organs present seven principal types of structure, distinguished from each other by the prominence of some particular part.

To make this clear, it may be as well to re-state that, in the mastax of a *Brachionus*, there are two hammer-like bodies or *mallei* (fig. 29, *ms*), which work on a kind of split

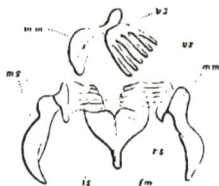

anvil or *incus* (fig. 29, *is*), and that each *malleus* consists of an upper part, the head or *uncus* (fig. 29, *us*), and of a lower part or handle, the *manubrium* (fig. 29, *mm*); while the *incus* consists also of two parts, the upper divided into two symmetrical halves, the *rami* (fig. 29, *rs*), which are supported on the lower piece or *fulcrum* (fig. 29, *fm*).

Now, in *Brachionus* all the trophi are well developed, but the other typical manducatory organs may be arranged in a series in which the *mallei* are successively degraded, while continually greater prominence is given to the *incus*; at least in all but three types; and in two of these the *rami* and *unci* are the prominent parts, while the third is distinguished by the close connection of the *mallei* and the *rami*.

Fig. 29.   Mallente.

The typical trophi may, then, be named as follows :

1. Malleate (fig. 29).

Mallei stout; manubria and unci of nearly equal length ; unci 5- to 7-toothed ; fulcrum short ; as in *Brachionus urceolaris*.

2. Sub-malleate (fig. 30).

Mallei slender ; manubria about twice as long as the unci ; unci 3- to 5-toothed ; as in *Euchlanis deflexa*.

3. Forcipate [1] (fig. 31).

Mallei rod-like ; manubria and fulcrum long ; unci pointed or evanescent ; rami much developed and used as a forceps ; as in *Diglena forcipata*.

_____
[1] In *Furcularia*, and in a few other genera, the rami as well as the mallei are rod-like ; and the

4. *Incudate* (fig. 32).

Mallei evanescent ; rami highly developed into a curved forceps ; fulcrum stout ; as in *Asplanchna Ebbesbornii.*

5. *Uncinate* (fig. 27).

Unci 2-toothed ; manubria evanescent ; incus slender ; as in *Stephanoceros Eichhornii.*

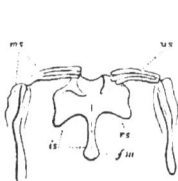

Fig. 30.—Sub-malleate.   Fig. 31,—Forcipate.   Fig. 32.—Incudate.

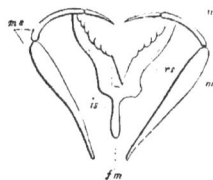

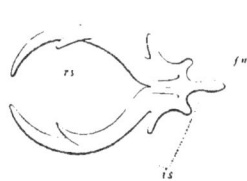

6. *Ramate* (fig. 28).

Rami sub-quadrantic, each crossed by two or three teeth ; manubria evanescent ; fulcrum rudimentary ; as in *Philodina roseola.*

7. *Malleo-ramate* (fig. 26).

Mallei fastened by unci to rami ; manubria 3 loops soldered to the unci ; unci 3-toothed ; rami large, with many striæ parallel to the teeth ; fulcrum slender ; as in *Melicerta ringens.*

Now, the seven Rotifera, made use of above to yield examples of typical trophi, are very distinct from each other, and show that the form of the trophi is one good characteristic for separating the families. But a difference in the shape and disposition of the corona, and of its ciliary wreath, generally accompanies a difference in the manducatory organs ; and the three together will serve as good guides to a division of the four orders into families.

In one of the sub-divisions of the *Ploïma*, however, the corona, ciliary wreath, and trophi are often difficult of determination ; but just where these guides desert us, a new character, viz. the lorica, comes to our aid, and shows such well-marked differences in shape and structure, as to enable us to divide this sub-order (the *Loricata*) into fairly natural groups. The *Loricata* are so called from the integument of the body ; which, from the distribution of chitine throughout the tissue, is hardened into a stiffened coat or shell (*lorica*, a coat of mail) inclosing, more or less completely, the internal organs. In the *Il-loricata* the integument is soft and flexible ; but there is, unfortunately, no very sharp division between the two sub-orders in this respect ; as every variety of integument exists, from the hard, dense coat of *Dinocharis*, to the tough yet flexible covering of *Rattulus*, and the perfectly soft cuticle of *Albertia*.

The following scheme, then, is an attempt to divide the four orders of Rotifera into families, by means of the various characters which I have just detailed : of course, there are some genera which do not readily fall into the arrangement ; but this is only what is certain to happen to every possible scheme of classification.

Such difficulties must attend every attempt to marshal Nature's endless varieties into

whole apparatus looks like one forceps within another. To this variety of the *forcipate* trophi the term *virgate* will be applied.

well-marked battalions. Nature knows no hard lines of separation; and the best of classifications can be only that which contains the fewest faults : as Müller has forcibly said, ' Optima est, quæ paucioribus horret.'

Order I. RHIZOTA.

Fixed when adult; foot transversely wrinkled, not retractile within the body, ending in a sucking disk or cup.

Fam. 1. *Flosculariadæ* (Pl. C, fig. 1).

Corona produced longitudinally into setigerous lobes ; *buccal orifice* central ; *ciliary wreath* a single half-circle above the buccal orifice ; *trophi* uncinate.[1]

Fam. 2. *Melicertadæ* (Pl. C, fig. 2).

Corona not produced into setigerous lobes ; *buccal orifice* lateral ; *ciliary wreath* a marginal continuous curve, bent on itself at the dorsal surface, so as to encircle the corona twice, with the buccal orifice between its upper and lower curves, and having also a dorsal gap between its points of flexure ; *trophi* malleo-ramate.[1]

Order II. BDELLOIDA.

That swim with their ciliary wreath, and creep like a leech ; foot wholly retractile within the body, telescopic, furcate.

Fam. 3. *Philodinadæ* (Pl. C, fig. 3).

* *Corona* two transverse circular lobes ; *ciliary wreath* a marginal continuous curve bent on itself at the dorsal surface, so as to encircle the corona twice, with the *buccal orifice* between its upper and lower curves, and having also two gaps, the one dorsal between its points of flexure, and the other a ventral gap in the upper curve opposite to the *buccal orifice* ; *trophi* ramate.[1]

Fam. 4. *Adinetadæ* (Pl. C, fig. 4).

Corona a flat ventrally placed surface ; *ciliary wreath* the furred ventral surface of the corona ; *trophi* ramate.[1]

Order III. PLOÏMA.

That swim with their ciliary wreath, and (in some cases) creep with their toes.

Sub-order. Il-loricata.

Foot, when present, almost invariably furcate ; but not transversely wrinkled ; rarely more than feebly telescopic, and partially retractile.

Fam. 5. *Microcodidæ*.

Corona obliquely transverse, flat, circular ; *buccal orifice* central ; *ciliary wreath* a marginal continuous curve encircling the corona, and two curves of larger cilia, one on each side of the buccal orifice ; *trophi* forcipate[1] ; *foot* stylate.

Fam. 6. *Asplanchnadæ* (Pl. C. fig. 7).

Corona two transverse, flattened, confluent cones, with their summits distinct ; *ciliary wreath* single, marginal ; *trophi* incudate[1] ; *intestine, cloaca,* and *foot,* absent.

[1] For description of these technical terms, see pp. 28, 29.

Fam. 7. *Synchætadæ* (Pl. C. fig. 6).

Corona a transverse spheroidal segment, sometimes much flattened, with styligerous prominences; *ciliary wreath* a single interrupted marginal curve, encircling the corona; *trophi* forcipate [1]; *foot* minute, furcate, or absent.

Fam. 8. *Triarthradæ.*

Corona transverse; *ciliary wreath* single, marginal, fringing the buccal orifice; *trophi* malleo-ramate [1]; *foot* absent.

Fam. 9. *Hydatinadæ* (Pl. C, fig. 5).

Corona truncate, with styligerous or ciliated prominences; *ciliary wreath* two parallel curves, the one marginal fringing the corona and buccal orifice, the other lying within the first, the styligerous prominences being between the two; *trophi* malleate [1]; *foot* furcate.

Fam. 10. *Notommatadæ.*

Corona obliquely transverse; *ciliary wreath* one of interrupted curves and clusters, usually with a marginal wreath surrounding the buccal orifice; *trophi* forcipate [1]; *foot* furcate.

Sub-order. Loricata.

Corona and *ciliary wreath* various in shape, but never Rhizotic, and Bdelloidic only in the *Pterodinadæ*; *trophi* of different types, but never Bdelloidic, and Rhizotic only in the *Pterodinadæ.*

Division I.

Foot jointed, stylate or furcate; not transversely wrinkled, nor wholly retractile.

Fam. 11. *Rattulidæ.*

Lorica entire, cylindrical, without angles; *trophi* asymmetrical.

Fam. 12. *Dinocharidæ.*

Lorica entire, vase-shaped, sometimes facetted; head distinct, with a chitinous covering; *trophi* symmetrical.

Fam. 13. *Salpinadæ.*

Lorica compressed, cleft down the back, the two halves united by a membrane, so as to form a dorsal furrow.

Fam. 14. *Euchlanidæ* (Pl. C. fig. 10).

Lorica depressed, of two dissimilar plates, one dorsal and one ventral, united by a membrane so as to form a lateral furrow.

Fam. 15. *Lepadelladæ.*

Lorica depressed, broad, closed beneath; head distinct, surmounted by a retractile, arched, chitinous plate.

Fam. 16. *Coluridæ.*

Lorica compressed, open beneath; head distinct, surmounted by a retractile, arched, chitinous plate.

[1] For description of these technical terms, see pp. 28, 29.

Division II.

Foot transversely wrinkled, wholly retractile, furcate or ending in a ciliated cup; sometimes absent.

Fam. 17. *Pterodinadæ* (Pl. C. fig. 9).

Lorica greatly depressed, entire, of two nearly equal plates, soldered together at the edges; *corona* and *ciliary wreath* those of the *Philodinadæ*: *trophi* malleo-ramate [1]; *foot* ending in a ciliated cup.

Fam. 18. *Brachionidæ* (Pl. C. fig. 8).

Lorica depressed, entire, dorsally arched, generally armed with anterior and posterior spines; *corona* transverse with styligerous prominences; *ciliary wreath* single, marginal, fringing the buccal orifice; *trophi* malleate [1]; *foot* furcate, or absent.

Order IV. Scirtopoda.

That swim with their ciliary wreath, and skip with Arthropodous limbs; foot *replaced by two dorsal, stylate, unconnected appendages, ending in ciliated expansions.*

Fam. 19. *Pedalionidæ* (Pl. C. fig. 11).

Corona truncate; *ciliary wreath* a marginal continuous curve, bent on itself at the dorsal surface, so as to encircle the corona twice, with the buccal orifice between its upper and lower curves; having also two gaps, the one dorsal between its points of flexure, and the other a ventral gap in the upper curve opposite to the buccal orifice; *trophi* malleo-ramate.[1]

The further subdivision of each family into genera will be given with the description of that family.

[1] For description of these technical terms, see pp. 28, 29.

CHAPTER IV.

ON THE HAUNTS AND HABITS OF THE ROTIFERA.

Nonne vides, quæcunque mora fluidoque calore
Corpora tabuerint, in parva animalia verti ?

Ovid, Metam. xv. 362.

Equidem tum Naturæ rerum gratias ago, cum illam non ab hac parte video,
quæ publica est, sed cum secretiora ejus intravi.

Curiosus spectator excutit singula et quærit. Quidni quærat ? Scit illa ad se
pertinere. Quantum enim est, quod ante pedes jacet.

Seneca, Nat. Quæst. præf. (adapted by O. F. Müller).

'T is born with all : the love of Nature's works
Is an ingredient in the compound man,'
Infused at the creation of the kind. . . .
It is a flame that dies not even there
Where nothing feeds it : neither business, crowds,
Nor habits of luxurious city life,
Whatever else they smother of true worth
In human bosoms, quench it or abate.

Cowper, The Task.

CHAPTER IV.

ON THE HAUNTS AND HABITS OF THE ROTIFERA.

THAT the first thing to be done is to catch your game, is a maxim as applicable to Rotifera as to hares; and it is no less true of these that to hunt for them successfully requires some knowledge of their haunts and habits. To carry away from a pond's side a bottle of dirty water full of slimy weed, is by no means a good plan for catching these animals, even the commonest and coarsest. It is true that there are some fine forms which may be found in very dirty ponds, or even in dirtier puddles: for instance, there can hardly be too dark a farmyard puddle for *Hydatina senta*, which rejoices in the drainings of a manure-heap, even when the water is of so deep a colour that it is impossible to see the animals in it when you have got them. *Triarthra*, too, and the beautiful *Notops* [1] *clavulata* are to be met with in cattle ponds, where the water is like pea-soup; and *Brachioni* of all kinds rejoice in such places, especially when green with *Euglenæ* and alive with the motile seeds of algæ. Indeed, there is one *Brachionus, B. angularis*, whose presence in a pond bids us put up our bottles and go elsewhere, as it likes water that will support hardly any Rotiferon but itself.

Floscularia, Stephanoceros, Melicerta, Limnias, and *Œcistes* are, of course, to be found only in such places as pond weeds will grow in healthily. Old ponds that have been left long undisturbed are their favourite haunts. *Floscularia* is a very wide-spread genus, at least so far as one or two species are concerned; and these may be looked for with every prospect of success in any such pond. Most of the finer and rarer kinds have been found in the Scotch lakes by Mr. Hood, who during the last four years has doubled the number of recorded species by his discoveries in the lochs round Dundee.

Stephanoceros, though by no means a rare Rotiferon, is more partially distributed; it is found often enough in ponds near London and Birmingham; but I have not heard that a single specimen has ever been met with in the neighbourhood of Clifton.

It appears also to be rare in Scotland; as Mr. Hood has found it only once or twice, in marsh-pools in Perthshire.

Melicerta ringens is to be found almost everywhere. It has even been seen swarming in one of the aquaria in the parrot-house in the Clifton Zoological Gardens. The roots of duckweed, the fibres of algæ, the leaves of *Myriophyllum*, and of all sorts of water plants, bear this very common species, as they do also the tubes of *Limnias* and *Œcistes*. *Lacinularia* and *Megalotrocha* have similar tastes, but are less frequently met with, especially the latter. This must be a comparatively rare genus, as it has been sent to me but three or four times in many years. *Cephalosiphon* is also rare. I once found a large colony of it, on a water weed at Nailsea in the big pond near the railway station, and it has been sent to me from Cheltenham and London; [2] but I never met with it again.

Conochilus is a lover of clear water. I have found it in Loch Lomond, and Dr. Imhoff has obtained it, in abundance, in the middle of Lake Zug. It is common enough

---

[1] *Notommata clavulata*; Ehr.

[2] Miss Saunders, to whom I am indebted for the specimens, says: "It is curious I never before came across this tube-dweller in the hundreds of pools I have searched."

in clean ponds round about London, and Mr. Hood has met with it in the Perthshire lochs.

The *Notommatadæ* and *Philodinadæ* have a very wide range. The algæ of ponds always hold many species, and many haunt the sediment that lies on the mud of the bottom. Some of the latter are to be found in gutters of houses, in water-butts, on the blades of wet grass, anywhere indeed where dust can fall and moisture can follow. For the eggs of the Rotifera are blown away by the wind from dried-up puddles, and are scattered broadcast through the air ; and some of the creatures they give birth to can exist apparently under almost any kind of moist conditions.

The *Asplanchnadæ* fortunately are as indifferent to their quarters as they are large and handsome. I have found them in roadside ditches thickly covered with *Lemna*, in farmyard ponds, in the clear water of a miniature lake, and in a foul yellow-green duck puddle in which the fluid (it could not be called water) was so thick that it had to be diluted with five or six times its own bulk, before anything could be seen in it. An *Asplanchna* is the very beau-ideal of a Rotiferon for a beginner. It is very large and transparent; it swims slowly ; and it generally occurs in great numbers. Moreover, its male is even more transparent than the female ; a mere living bubble, thinner and clearer than the finest blown glass. Unhappily, they are as capricious as they are charming; for a pool may be full of them to-day and deserted to-morrow ; and, so far as my own experience goes, they do not occur in the same spot year after year as many Rotifera do.

The *Pterodinadæ* and *Euchlanidæ* are dwellers in clear ponds, and rather solitary in their habits. On a warm sunny day the latter may be captured by skimming off the *Lemna*, and floating bits of leaves and stems, that are driven to the leeward corner of a pool. If the bottle be then allowed to stand a few minutes till the water is clear, a *Euchlanis* will often be found slowly gliding up the glass with its long toes pressed against it. It is always worth while to capture it at once with a pipette, and put it into a small tube along with any others of the same kind, as a live *Euchlanis*, properly exhibited under dark-field illumination, is one of the choicest treats that the Rotifera afford.

The *Pterodinadæ* are almost always creeping about the sides of the pond. I never but once have seen any other than solitary specimens, but that exception was a notable one. I then found swarms of them in the small space in which the sluice gate of a pond worked. It was not more than a foot square by about four feet deep, and was mantled over by duckweed. Out of this unlikely spot they were dipped by hundreds, while not one could be found in the pond itself. Of course the duckweed came up with the *Pterodinæ*, though as little as possible was taken, on account of the disagreeable way in which it clings to every pipette put into the bottle. On this occasion, however, I found it of great service ; for, on inspecting the catch at home with a hand-lens, I noticed that the Rotifera were attached in clusters to some of the roots of the duckweed. Watching for a favourable opportunity, I whipped one of the *Lemnæ* out of the water so suddenly, that the creatures had not time to let go their hold. Then cutting off the green head, I coiled the stem into a circle on a glass slip, and covered it with thin glass. It was impossible to conceive a more beautiful sight than this natural cage now afforded me. Thirty or forty shields of living glass were flashing across the field of view in every direction ; some were adhering to the stem, swaying backwards and forwards so as to present themselves in every point of view, while others were moored to the glass cover, thus giving an admirable opportunity of making out their structure.

It was a memorable occasion, but I never had such a chance again.

I have yet to speak of the *Synchætadæ* among the free-swimmers. Both the genera included in this family are to be found in open water ; and both alike shun dirty ponds ; though in different degrees, for *Synchæta* is absolutely intolerant of them ; while I have dipped up *Polyarthra* from the hollows of a muddy bottom where once a pond had been.

Both genera are tolerably common, and are often to be caught in considerable numbers. The habits of the two chief species of *Synchæta*, viz. *S. mordax* and *S.*

tremula, are very different. The former is the swiftest and most restless of the Rotifera ; it is the very swallow of the waters, ever whirling round and round in endless spirals, and never still for a single instant from its birth till its death ; but the latter may be constantly seen drifting along in some gentle current, while twisting round at the end of a long thread spun from its toes, and fastened to some floating object.

Of course, creatures with habits like these can be captured only by making random dips in the water, now at the surface, now deeper down—here in the sunshine, and there in the shade ; for even Rotifera have their fancies, and are sometimes swarming in one particular spot, while all the rest of the pond is deserted by them.

I have, however, noticed that they specially affect the neighbourhood of a forest of weeds growing up from the bottom ; waltzing up and down outside of them in myriads, like gnats under the trees in summer.

There is yet another free swimmer that avoids the shore, and sails out into the open sea ; viz. *Anuræa longispina*. This curious creature has a lorica like a Greenlander's canoe, or a University eight, and it keeps off from the weeds and algæ, as if fearing lest it should be entangled for life if it once got among them. It was discovered by Professor D. S. Kellicott in Niagara water at Buffalo, U.S., in 1879, and was found almost immediately afterwards in the Olton Reservoir, near Birmingham and since then in Lake Zug in Switzerland. It appears to be a rare species, though its rarity may be due partly to the fact that it often requires a boat to catch it ; and an ordinary Rotifer hunter can hardly be expected to add this to his apparatus.

The known habitats of the *Pedalionidæ* are at present very few. I had the good fortune to be the first to light on *Pedalion mirum*. It was in July 1871 that I found it in a small roadside pool at the top of Nightingale Valley, close to Clifton. Soon afterwards I dipped it from a fine old pond at Abbot's Leigh, about two miles distant from Nightingale Valley. It reappeared in this pond in the following year, but since then it has not revisited the neighbourhood. It has been met with several times near Chester and Birmingham, and on one occasion it was tolerably abundant in the warm water-lily tank in the Duke of Westminster's gardens at Eaton.[1]

Dr. L. K. Schmarda discovered in Egypt, in 1853, in some brackish pools near El-Kab, a six-limbed Rotiferon, *Hexarthra polyptera*, which evidently belongs to the same family, though it must be placed in a different genus. He says that there were great swarms of them distinctly visible to the naked eye, in a pool of very transparent, colourless water, of a strong brackish taste.

Now, a Rotiferon that is equally at home in dirty puddles, clear ponds, warm-water tanks, and brackish pools, ought not to be a rare one : and yet *Pedalion is* rare.

Possibly its apparent rarity is due to its being constantly mistaken for an Entomostracous larva. I was on the point of throwing the water away, when I first dipped *Pedalion* out of the pool in Nightingale Valley. Its skipping movement is so precisely that of the young of a *Cyclops*, that I thought I had caught nothing more valuable than these ever-present nuisances. Fortunately I noticed that, unlike them, my captives seemed to glide along after every skip, instead of stopping stock-still to gather breath for a fresh jump ; and so, thinking that they might possibly be some large sort of *Polyarthra*, took them home for further investigation. But it is very probable that *Pedalion* has been thrown away hundreds of times, and will be so again, as this happened to me after nearly twenty years' experience in catching Rotifera.[2]

Rotifera may often be seen perched just under the plumed heads of one of the fresh-

---

[1] M. J. Barrois described, in the *Revue Scientifique*, No. 13, 1877, p. 303, a marine Rotiferon under the name *Pedalion*, and gave an account of its embryology. His description, however, shows that the animal was of the genus *Synchæta*.

[2] It is a pity that *Pedalion* is not more frequently met with, as there are some points in its structure that yet remain to be cleared up ; and as it is such a striking link between the *Rotifera* and the *Arthropoda*. Mr. T. Bolton, of Birmingham, has, however, succeeded in preserving specimens as microscopic objects, and they can generally be obtained from him.

water zoophytes, wisely making use of the stronger currents produced by the ciliated tentacles of their hosts, in order to bring grist more easily to their own mills. I have had sometimes quite "a happy family" of them in the field of view at once; a *Brachionus*, a *Philodina* or two, and a *Melicerta*, all attached to the neck of a *Plumatella*, and all eagerly whirling their wheels in order to divert to their own throats a portion of the currents that swept down to them from above. Nor was this all; for the *Melicerta* in its turn had the top of its tube turned to the same use, and bore, as closely under its wheels as possible, the tiny case of one of its own offspring.

Limnias ceratophylli and *Melicerta ringens* carry this semi-parasitical habit to a great extent. Clusters of two or three generations all attached to one tube are not at all uncommon in the former species; and I once found in Nailsea pond a large *Limnias* bearing up no fewer than fourteen of its own descendants. *Melicerta ringens*, too, in America,[1] is frequently met with in large adhering clusters, but in England it is usually a solitary species. However, this is not invariably the case; for not long ago I had the pleasure of seeing as many as thirty-four live *Melicertæ* attached to one another.[2] They were of all ages and sizes, and were grouped round one large tube, so as to form a striking example of a natural co-operative society. Nor is this the only way in which the Rotifera show their capacity for fighting the battle of life. Every animal is limited by its own powers to a certain space, beyond which its excursions cannot possibly extend. Its food and its mate must be found within these limits; and when these two imperious wants are satisfied, there is but little time or strength left for travelling. But it would be an obvious advantage to many creatures if they could be carried about from one spot to another without tiring their own muscles—ready to slip off, at any favourable opportunity, "to fresh woods and pastures new." Now this is precisely what some of the *Brachioni* and *Philodinæ* contrive to do; for they may be seen riding in clusters on the backs and sides of the Entomostraca, or thickly fringing the legs and side plates of the water wood-louse.[3]

Whenever I have caught a water-flea[4] so encumbered, and have placed it in a live box to see the Rotifera it carried, they have soon deserted their captive steed, and have swum off as if to search for a more serviceable one.

There are, too, some Rotifera whose structure has been adapted to give them a good grasp of their host, or even to enable them to pierce its skin, and so suck its juices for their own support.

Balatro calvus, for example, has been found[5] in the Seine (Canton de Genève) creeping on the bodies of small water-worms which it habitually infests, and having two greatly enlarged foot-processes, which probably enable it to take a firm hold.

Another Rotiferon, *Callidina parasitica*, is always found attached to the thoracic or abdominal appendages of the fresh-water shrimp[6] and water wood-louse, and limits its journeyings to creeping about on the body of its host; while the strange creature *Drilophaga Bucephalus* holds on by its altered jaws to the hind segments of a fresh-water worm, *Lumbriculus variegatus*, and sucks the animal it clings to.

This parasitic Rotiferon was discovered in North Bohemia in the great pond at Hirse-berg, in the banks of which the *Lumbriculus* is found in immense numbers. The worm buries the fore part of its body in the mud, and moves its naked hinder segments like a pendulum in the water. But no such gentle motion will unfasten the grip of the *Drilophaga*, which is so firmly attached to the worm's skin that it can be dislodged only by using considerable force.

[1] I am indebted to Mr. Galloway C. Morris, of Philadelphia, U.S., for this information. A cluster of tubes that he sent to me contains twenty-eight specimens of various sizes adhering together.

[2] It was found near Clifton by Mr. E. C. Bousfield, and is drawn in Pl. V. fig. 1, c.

[3] *Asellus vulgaris.*　　[4] *Daphnia pulex.*

[5] By M. Ed. Claparède.　See *Ann. Sci. Nat. Zool.* 5 Ser. vol. viii, 1867.

[6] *Gammarus pulex.* Mr. H. Giglioli, who discovered this species, says that he has never found it anywhere else; and that out of 700 *Gammari* from different localities, not one was free from the parasite. *Quart. J. Micr. Sci.* N. Ser. vol. iii, 1863, p. 237.

If the Rotifer hunter can rise to a pitch of enthusiasm, which I confess I have never been able to attain to, he may follow Dujardin's example, and, by making incisions [1] in the sides of earth-worms and slugs, obtain from the expressed fluids the entozoic *Albertia vermiculus.* The same creature has been seen inside *Nais proboscidea,*[2] and an allied species (*Albertia crystallina*) in the viscera of *Nais littoralis,*[3] while the *Synapta* of the Channel Islands have been found[4] to carry within their body-cavity a minute Rotiferon only $\frac{1}{500}$ of an inch in length.

But I have not yet exhausted the list of these strange dwelling-places. A pretty little Rotiferon, *Notommata parasita,* may be found swimming about within the beautiful spheres of *Volvox globator,* or lodged within the embryo globes when almost ready to escape from the parent sphere. "On examining several specimens of *Volvox* with a pocket lens we may frequently detect such as are thus tenanted, by perceiving a spot differing from the young clusters in form and colour. These spots are found to be the *Notommata,* snugly ensconced within the globe, in the spacious area of which it lives at ease, and swims to and fro like a gold fish in a glass vase. We see it for the most part, however, clinging to the inner surface of the circumference, engaged in devouring the green monads with which the gelatinous surface is studded, or else eating away the embryo clusters."[5] *Volvox globator* is common enough in most neighbourhoods, and may be met with it even in clear rain puddles in quarries and plantations; while in two or three ponds near Clifton it is sometimes so abundant as to give a green hue to the water; and yet I have never seen its guest here, nor do I know anyone in the neighbourhood who has; so it can scarcely be a common species.

The reproductive cells of *Vaucheria*—a thread-like alga which grows on pond walls and in many moist places—are the homes of another *Notommata, N. Werneckii.* This parasitic Rotiferon passes a small portion of its youth in the open water; but it soon returns to a lifelong imprisonment in the green cells in which it was hatched, and where it undergoes very singular changes of form. Its presence in the *Vaucheria* may be detected by the unusual size and shape of the reproductive cells, and by their containing a black spot which is the animal's stomach.[6]

Some further means of obtaining Rotifera have yet to be mentioned.

If a little of the mud or rotten leaves at the bottom of a dried-up pool, in which Rotifera have been observed, is brought home and allowed to lie in a vessel of water, the chances are in favour of there being in the mud some of their eggs, and of their ultimately hatching. I have often adopted this plan with success, especially when some rare species has been discovered in a little pool due only to the rain, and drying up after two or three days' fine weather. Unluckily the mud too frequently harbours an abundance of small worms also; and these are disagreeable to see and troublesome to deal with, for they are liable to starve, die, and taint the water.

Rotifera also may be produced at home by placing infusions of hay, leaves, &c. in some vessel out of doors. No very great variety is to be obtained by such methods; but it is always as well to try it, and to have a good-sized pan in the garden, full of soft water, into which rubbish from pond-gatherings may occasionally be thrown. These, and the chance droppings from the air into the pan, will sometimes give the student, at his own door, species which he would otherwise have to travel far to find.

Many of the Rotifera may be kept indoors in vessels in which there is a healthy growth of *Myriophyllum, Anacharis,* or other water-weed. Mr. Gosse has tried this plan with

---

[1] *Ann. Sci. Nat. Zool.* 2 Ser. vol. x. 1838, p. 176.
[2] By Mr. P. H. Gosse, in water from a pond at Walthamstow.   [3] By M. Max Schulze.
[4] By Professor Ray Lankester, *Quart. J. Micr. Sci.* N. Ser. vol. viii. 1868, p. 51.
[5] Mr. P. H. Gosse, *Trans. Micr. Soc.* vol. iii. 1852, p. 113.
[6] Prof. Balbiani, *Ann. Sci. Nat.* 6 Ser. vol. vii. 1878. I am not aware that *Notommata Werneckii* has been found in England. Probably it would be, were it deliberately searched for. Mr. F. W. Roper, of Eastbourne, has found a similar species tightly rolled up in a ball inside the leaves of one of the liver-worts, *Lejeunia minutissima.*

success, and has lately had thousands of *Stephanoceros*, *Melicerta*, *Pterodina*, &c., thriving in tanks and jars in his study. Mr. J. Hood has been equally successful with *Floscularia* and *Limnias*; and I know of other instances in which a literally constant stock of the tube-making Rotifera has been maintained in these home preserves.

Indeed, if nothing more is desired than to watch the growth of a couple of generations or so, an ordinary zoophyte trough is aquarium enough. All that is necessary is, (1) to take great care that there are not many animals of any kind in it, (2) to keep it in a subdued light, (3) and at a moderate temperature, (4) and especially to provide the Rotifera with plenty of their natural food. For, in the great majority of cases, Rotifera die, when in captivity, of starvation: one moment's examination of their stomachs will make that point clear.

Of course, what is their natural food must first be observed under the microscope, and then it must be provided to them every day by dropping a fresh supply of water containing it into the trough. It will, therefore, always be necessary to bring away from the pond, where they were found, a good supply of pond water free from all other kinds of animals.

Sometimes, however, it is well to make an overfed Rotiferon starve a little, in order to see its internal structure. *Hydatina senta*, for instance, is frequently so gorged with dark green food, that little else can be seen but its distended stomach; the organization of *Pedalion mirum*, too, is often a hopeless riddle, owing to its greedy habits; but drop either of these creatures into a tube of clear soft water for an hour or so, and it may be fetched out again in delightful condition for microscopic investigation, and yet perfectly healthy.

Summing up the various habitats that I have just recorded, we see that Rotifera may be found in rivers, lakes, reservoirs, ponds, ditches, puddles, gutters, and water-butts; in the mud of dried ponds, in the dust of dried house-gutters, on wet moss and grass; in the rolled-up leaves of liver-worts, in the cells of *Volvox globator* and of *Vaucheria*, in vegetable infusions; on the backs of *Entomostraca*, and of fresh-water fleas, wood-lice, shrimps, and worms; in the viscera of slugs, earth-worms, and *Naiades*, and in the body-cavities of *Synapta*. Nor have I yet completed the list; for several species have been found in the sea. Mr. Gosse says,[1] "*Synchæta Baltica* swims at large through the water, never resting; it is self-luminous, and is one of the causes of the phosphorence of the sea. *Brachionus Mülleri* and *Pterodina clupeata* occur in brackish water at the mouths of rivers; and other marine species may often be detected by searching with a pocket-lens the glass sides of a well-stocked aquarium."

Since then these creatures have so wide a range of habitats, it is hardly possible for anyone, who will take the trouble, not to find some of them near his own home.

To obtain some particular Rotiferon, at a particular given time, is often difficult enough, if not impossible; but for one who is content to study these beautiful creatures as he finds them, there is always a never-ending supply of delightful amusement.

[1] *Marine Zoology*. part i. 1855. p. 107. See also *Tenby*, 1856, p. 271.

EXPLANATION OF THE LETTERING OF THE PLATES.

. Corona . . . .
. Lobe of do. . . .
. Styligerous prominence .
. Chin . . . .
. Ciliated cup . . .
. Hook . . . . } Parts of head and body.
. Horny processes . .
. Foot . . . .
. Toe . . . .
. Ciliary wreath . .
. Principal wreath . .
. Secondary wreath . .
. Taster . . . .
. Vestibule . . .
. Buccal funnel . . .
. Lips . . . .
. Tube from lips to crop .
. Crop . . . .
. Mastax . . . .

Trophi { us } . malleus { mm } manubrium   Nutritive system.
or { us } mucus
Jaws { is } . incus { rs } ramus
{ fu } fulcrum

. Œsophagus .
. Stomach .
. Intestine .
. Rectum . .
. Cloaca . .
. Salivary gland . } Secreting system.
. Gastric gland . .
. Foot gland . .
. Lateral canal . } Vascular system.
. Vibratile tag .
. Contractile vesicle .
. Ovary . . .
. Germ . . .
. Ovum . . } Reproductive system.
. Oviduct . .
. Penis . .
. Sperm-sac . .
. Longitudinal muscle } Muscular system.
. Transverse muscle .
. Ganglion . .
. Eye . . .
. Dorsal antenna . } Nervous system.
. Lateral antenna .
. Tactile style . .
. Nerve thread . .

EXPLANATION OF PLATE A.

. *In all case. throughout the plates, both the plain and coloured, the figures that have the initial H attached to them have been drawn from life by Dr. Hudson, and those with the initial G, also from life, by Mr. Gosse.*

BRACHIONUS RUBENS *and the details of its structure.*

1. Dorsal view	. H	9. Mallens. as in nature	.	. G
2. Ventral view; upper half .	. H	10. Incus, side view	.	. G
3. Lorica and muscles; dorsal view	. H	11. Ganglion and eye .	.	. H
4. „ „ ventral view	. H	12. Extremity of foot .	.	. H
5. Side view	. H	13. Male; penis withdrawn (after Dr. Cohn)		
6. Mastax and trophi; dorsal view .	. G	14. Male; penis extruded (after Dr. Cohn)		
7. Trophi; ventral view . .	. G	15. Dorsal antenna . . .	.	. H
8. Mallens expanded by pressure .	. G	16. Ephippial egg (after Dr. Cohn)		

In all the above figures the letters signify as follows :—

sp .	. styligerous prominence	*cl* .	. cloaca	
f .	. foot	*sy* .	. salivary gland	
t .	. toe	*gg* .	. gastric gland	
cw .	. ciliary wreath	*fg* .	. foot gland	
bf .	. buccal funnel	*lc* .	. lateral canal	
lp .	. lips	*vt* .	. vibratile tag	
mx .	. mastax	*cv* .	. contractile vesicle	
ti .	. trophi	*oy* .	. ovary	
mx .	. mallens	*g* .	. germ	
us .	. uncus	*om*	. ovum	
mm .	. manubrium	*p* .	. penis	
is .	. incus	*ss* .	. sperm sac	
rs .	. ramus	*lm* .	. longitudinal muscle	
fm .	. fulcrum	*gn* .	. ganglion	
o .	. œsophagus	*r* .	. eye	
s .	. stomach	*a* .	. dorsal antenna	
	intestine	*a'* .	. lateral antenna	
	rectum	*ts* .	. tactile style	

Pl. A

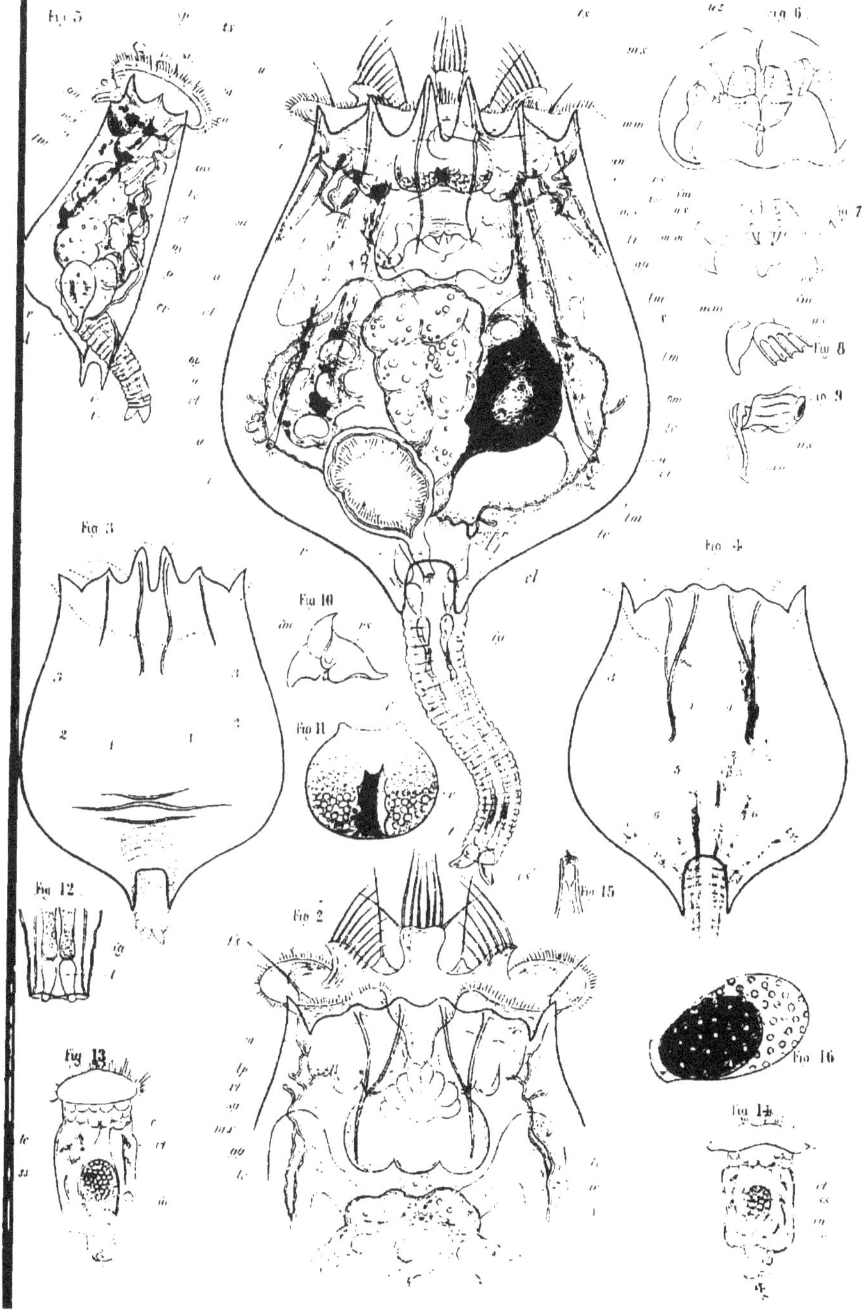

Fig 1
Fig 2
Fig 3
Fig 4
Fig 5
Fig 6
Fig 7
Fig 8
Fig 9
Fig 10
Fig 11
Fig 12
Fig 13
Fig 14
Fig 15
Fig 16

EXPLANATION OF PLATE B.

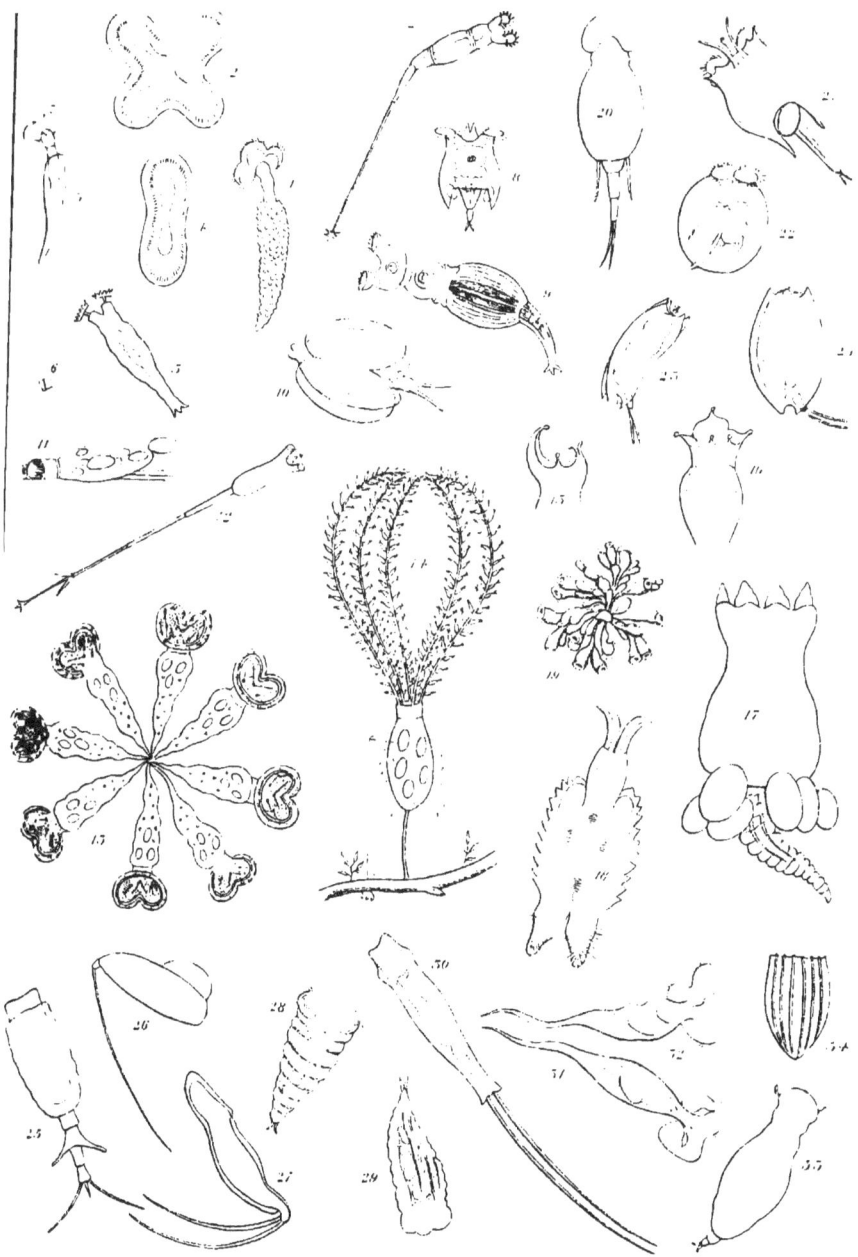

PLATE I.

**** In all cases throughout the plates, where the sex is not mentioned, the animal is a female.*

1.	Floscularia campanulata	side view	H
1*a*.	" "	obliquely ventral	H
1*b*.	" "	male, same scale as 1	H
1*c*.	" "	male, enlarged	H
1*d*.	" "	female, trochal disc from above	H
2.	Floscularia ambigua	side view	H
2*a*.	" "	obliquely ventral	H
3.	Floscularia algicola	side view	H
4.	Floscularia longicaudata	dorsal view	H
5.	Floscularia coronetta	side view	H
5*a*.	" " (after Cubitt)	trochal disc, from above	
6.	Floscularia Cyclops	side view	H
7.	Floscularia cornuta	side view	H
8.	Floscularia regalis	side view	H
8*a*.	" "	dorsal view	H
9.	Floscularia ornata	obliquely dorsal	H
9*a*.	" " trophi	side view	G
9*b*.	" " "	obliquely from above	G
9*c*.	" " young female just hatched		G
9*d*.	" " the same a little older		G

PLATE II

PLATE III.

PLATE IV.

Details of Stephanoceros Eichhornii.

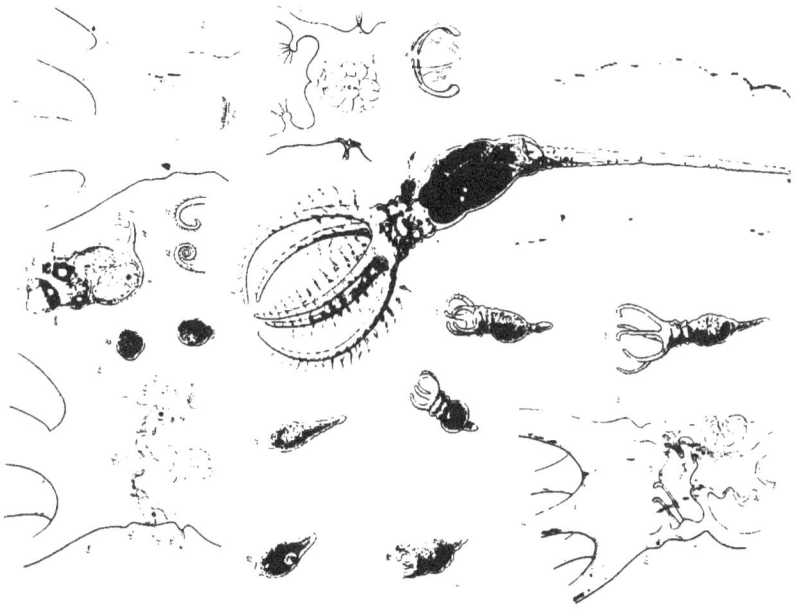

PLATE V.

CHAPTER V.

FLOSCULARIADÆ.

Difficultates, quibus laborat investigatio animalculorum microscopicorum, innumerae; eorundemque certa et distincta determinatio tantum temporis, tantum oculorum judiciique acumen, tantamque animi compositi et patientissimi praesentiam requirunt, ut vix aliud supra. Nihil facilius quam animalcula videre, eorumque motu et ludo delectari; differentias vero in bestiolis simplicissimis, mobilissimis, mutabilibus, in area minimi campi conspectum omni momento effugientibus, percipere, perceptas variosque cujusvis motus verbis significantibus exprimere, hic labor, hoc opus.

Hinc saepe post lucubrationem plurium horarum, cum videre et mirari lassus essem, defectu tamen verborum insolitos motus et imagines exprimentium, metuique, ne quae ipse quidem oculo et mente percepi lectori obscura manerent, chartae nihil commisi.   O. F. Müller.

CHAPTER V.

Order I. **RHIZOTA.**

Fixed when adult, usually inhabiting a gelatinous tube excreted from the skin; foot transversely wrinkled, not retractile within the body, ending in an adhesive disk or cup.

Family I. FLOSCULARIADÆ.

Coroua *produced longitudinally into setigerous lobes;* **buccal orifice** *central;* ciliary **wreath** *a single half circle above the buccal orifice;* **trophi** *uncinate.*

This family, like the one that follows it, contains some of the largest, handsomest, and most attractive of the Rotifera. It consists of only two genera, *Floscularia*, and *Stephanoceros*, which closely resemble each other in their habits and internal structure, but differ considerably in outward form. The latter genus, which is represented by only a single species, has its frontal lobes produced into long arms, having setæ set round them in whorls; while the former, which contains no fewer than sixteen species, has the lobes comparatively short and expanded, with the setæ radiating from their summits, and frequently edging the whole rim of the corona. The two genera differ also in the kind of tubes that they secrete. The majority of the Floscules have some what irregular tubes of slight consistency; but *Stephanoceros* has a thicker tube, more regular in shape, and apparently of greater density.

Both genera are to be found adhering to the common water-plants, frequently in the axils of the leaves; though some species prefer more exposed positions on the stems, or on the leaves themselves. The *Flosculariadæ* bear captivity fairly well, and may be easily bred, provided that they have a good-sized trough and a plentiful supply of food; for they are greedy feeders. They live mainly on *Monads*, ciliated *Protozoa*, &c.; but occasionally capture and swallow comparatively large animals, such as *Stentors*, or even free swimming *Rotifera*. In one respect, however, captivity often tells on them; for the home-bred specimens, though healthy enough, and breeding freely, are frequently much inferior in size to those that are brought fresh from their native haunts.

Genus FLOSCULARIA, *Oken.*

GENERIC CHARACTERS.—**Frontal lobes** *short, expanded, or wholly wanting;* **setæ** *very long and radiating, or short and cilia-like;* **foot** *terminated by a non-retractile peduncle, ending in an adhesive disk.*

Neither pen nor pencil can do justice to the beauty of these animated flowers. It can only be properly appreciated when they are seen by dark-field illumination under the microscope. Then the eye is at once delighted with the filmy transparency of the petaloid head, with the flowing curves of the lobes, and with the pencils of delicate setæ radiating from their summits in all directions, and often passing altogether out of the field of view. Should, moreover, the species under observation happen to be a social one, such as *F. campanulata*, four or five specimens may often be found with their tubes in juxtaposition, and the whole group can then be well shown under a low power; the animals, in various positions and under different aspects, forming, with their delicate cases and interlacing setæ, a picture that can be hardly rivalled.

F. campanulata, when fully expanded, has been compared to "a long tubular flower with a five-angled petal, the tube swollen, contracted below the lip, and seated on the end of a long stalk." [1] This description applies very well in most respects to the other species, except that the number of petals is not always five; for, owing to late discoveries, there is now a regular series of Floscules with seven, five, three, and two lobes; and one species in which the corona is not divided into lobes at all.

The setæ also, which crown the lobes, and are so highly characteristic of the better-known forms, vary quite as much in the newer species as do the coronæ; in some exceeding the Rotiferon's utmost length, and in others diminishing almost to the size of ordinary cilia. Indeed, if the strange genus *Acyclus*—which must be very closely allied to the Floscules—be also taken into account, as well as the equally curious genus *Apsilus*, there is a tolerably complete series of forms showing a gradual change from a Floscule, with seven lobes, and long radiating setæ, to a Floscule-like Rotiferon in which the setæ have entirely vanished, the corona has degenerated into a very delicate protrusile cup, and even the foot itself has shrunk into a mere sucking disk.

The **Tube.** The Floscules inhabit a semi-transparent gelatinous tube, into which the animal when alarmed can contract itself with great swiftness. It is secreted by the creature itself, and moulded on its own body by its sudden contractions, and slow expansions. When free from diatoms and extraneous particles (which is seldom the case), it is difficult to be seen, especially by transmitted light : under the dark-field illumination, not only can its outline be seen, but the substance of which it is composed can be traced from the outer surface, far in towards the Rotiferon itself. The tube becomes thinner towards the top, and it is often difficult to trace it there ; but it will generally be found to close in neatly a little under the neck. [2]

A Floscule, emerging from its tube, after one of its contractions, presents the appearance of a pear-shaped body on a transversely wrinkled stalk, with a pencil of long parallel hairs rising from the puckered centre of the rounded upper end. It slowly stretches itself till the wrinkles of the foot have nearly disappeared ; and then, after a delay, sometimes provokingly long, the puckers round the setæ relax, and the whole pencil is thrust forward, by the unfolding of the lobes of the corona ; which, as they rise, show that they had been drawn down into the body by inversion, as the tip of the finger of a glove may be drawn into it, by pulling it from within. After a little further hesitation, the lobes unfold, and expand into a wide cup, while "the setæ seem to fall round it on all sides in a graceful shower." The now fully expanded Floscule consists of five well-marked portions ; the corona, the vestibule, the crop, the trunk (including the viscera), and the foot.

The **corona** is a delicate nearly hemispherical cup, whose free edge is cut into lobes varying much in size, shape, and number. There are two main varieties of lobe ; in the one they are narrow, pointed, and ending in a spherical knob; in the other they are broad, bounded by low convex curves, and knobless. In almost all, the dorsal lobe is conspicuous by its greater size, or peculiar appendages. The corona is furled by the action of delicate muscular threads imbedded in its surface, and expanded by the upward rush of fluid between its outer and inner integuments, due to the contraction of the transverse muscles of the trunk.

The **setæ** are set either on the knobs that crown the summits of the lobes, or on a thick rim running round them ; and they sometimes form a continuous fringe on the

[1] Gosse, *Popular Sci. Rev.* vol. i. 1862, p. 160.
[2] Though the tube is of the flimsiest material, and lighter than water, it certainly protects the animal from its enemies. I have watched a sharp-jawed larva trying to bite through a Floscule's tube, and it was as completely foiled by its swaying about from side to side at every touch, as a boy at Hallow E'en, baffled by a floating apple, when trying to seize it with his teeth.

rim of the corona. Owing to their great delicacy, and to their lying in different planes, it is impossible to see nearly all of them at once. They vary greatly in size, position and arrangement; but their varieties, with those of the forms of the lobes, will be described in the account of each species.

Volvocina, small *Infusoria*, and floating particles may constantly be seen to enter the bell-shaped corona, and to pass thence down towards the buccal orifice. The setæ take no part in this process, beyond that of preventing the return of the captured prey, by interlacing in a close network over the top of the cup, or by individually lashing at a returning object, so as to throw it back again into the gulf. The interlacing of the setæ is accomplished by the heads of the lobes approaching each other, and, should the prey be large and vigorous, the lobes are pressed tightly together, so as to completely bar all chance of escape. In most of the species, the motion of the setæ appears due to the fitful action of the cuticle, on which they are placed; but in *F. trilobata, F. Hoodii,* and notably in *F. mutabilis,* a regular cilia-like motion occurs in the setæ; while in *F. mira* there is a still wider departure from the ordinary type; since in this Floscule each seta has a constant, slow, independent, amœboid motion.

The **Vestibule.**—At the bottom of the corona is a second chamber (the vestibule), bounded above by a highly contractile collar, below by a diaphragm with a slit in its centre (the buccal orifice), and on the sides by thick walls. On the upper margin of the collar, and running half round it on the ventral side, is a horse-shoe-shaped ciliated rim, ending in two knobs, bearing long, slowly moving cilia; and this rim is so set, that it slopes downwards from the dorsal side to the ventral. This true rotatory apparatus may be easily seen in the large Floscules *F. Hoodii* and *F. trilobata.*[1]

A current, due to the action of these cilia, sets down the coronal cup, in a plane at right angles to its base, and carries the food, past the collar, into the vestibule. When once an organism has reached the vestibule, there is no return for it to the upper world. The Floscule often suffers two or three small *Infusoria* &c. to wander about round the walls of the vestibule; but at any attempt to pass the collar, that at once contracts on itself, and closes the passage. In the diaphragm, which is the base of the vestibule, there is a long slit, the **buccal orifice,** bounded by two chitinous lips (Pl. I. fig. 1*d, lp*), from which there hangs into the next chamber, called the "crop," an elastic tube (Pl. II. fig. 4, *t*), which may be seen always undulating above the mastax. When there are victims enough collected in the vestibule to make it worth while to swallow them, the collar contracts violently, the lips dart forward with a sort of snap, and the prey is forced down the tube into the crop. It is evident that this hanging tube is an admirable contrivance for admitting fresh prey into the crop, while at the same time preventing the return of that previously captured. Naturalists plagiarise from the Floscules, when they drop their live specimens through a quill stuck into the cork of a bottle; only the rigid quill is far inferior to the flexible and ever-moving tube.

The **crop** (Pl. II. fig. 4, *cp*) is a rounded chamber just under the diaphragm at the base of the vestibule. It has very thick walls, which are strengthened externally by two granular spots one on either side of the Floscule's shoulders (Pl. I. figs. 1, 8*a*). Viewed as opaque objects they are white, like the similar oval knobs on *Megalotrocha alboflavicans.*

Under the action of small muscular fibres, the sides of the crop contract alternately, and throw the contained food from side to side; by which means every part of it in turn is subjected to the action of the jaws (Pl. II. fig. 1, *ti*). These lie at the bottom of the

[1] Dr. Dobie described the two ciliated knobs in *F. campanulata, Ann. Nat. Hist.* 1849. Mr. Gosse inferred the existence and position of the true rotatory organ from the motion of particles in the coronal cup (*Tenby,* p. 307). Dr. Moxon says that "the alimentary canal above the gizzard is divided by a highly irritable cilium-clothed sphincter of irregular outline" (*Trans. Linn. Soc.* vol. xxiv. 1864, p. 457). In 1867 I published a full description of the rotatory organ of *F. campanulata* in the *Trans. Bristol Micr. Soc.* In 1869 Mr. Cubitt did the same for *Stephanoceros* and *F. coronetta* (*Mon. Micr. J.* vol. ii. 1869, p. 133).

crop, attached to the walls of the stomach ; and to reach the latter everything must pass between them. The Floscules are great feeders, and sometimes the crop becomes so distended with food, that the animal, unable to force it past the jaws quickly enough, seeks relief by expelling the contents of the crop right through the inverted tube. The lobes of the corona are folded back on the body, the diaphragm is pushed upwards and the tube is thrust inside out through the slit in the diaphragm ; while through it pours the unmanageable food. Dr. Moxon (loc. cit.) has seen this take place on two separate occasions, and I have seen it once : in each case the animal was *F. campanulata.*[1] The appearances due to the tube have been variously described as caused by moving filaments, laminae, vibratile cilia, and a waving membrane ; but these observations of Dr. Moxon, confirmed by my own, put the matter beyond doubt.

The **Trunk.** The outer wall of the trunk is a tough, elastic, and often shining cuticle, which has an inner and softer layer of varying thickness. This double covering interferes greatly with a clear view of the viscera, especially as it has intercommunicating cavities and channels containing fluid, which is driven upwards and downwards by the contraction of the muscles, and by the various motions of the body. Nor is this all ; for the fluid itself is often rendered semiopaque by granules floating in it. It is doubtless by means of this fluid that the lobes of the furled corona are pushed forward and expanded, the transverse muscles of the trunk forcing it into definite channels, which are thus rendered tight and stiff, like the ribs of an umbrella. Mr. Gosse[2] has described and figured these in the case of *F. ornata* ; and has noticed how the granules flow from the trunk over the neck into the various channels of the coronal cup. Mr. Hood, too, has watched a steady stream of granules passing down from the trunk into the foot, and returning again from a point about half way from its extremity. The granules may be frequently seen, in some degree, in specimens of almost every species, but occasionally they are in such abundance as to render the animal quite opaque ; and, by reflected light, of a dead white.

The **foot** is very long and flexible, and is capable of great expansion and contraction, but cannot be drawn into the trunk. It consists of little else but muscles. The great longitudinal muscles pass down its whole length, and numerous fine muscular fibres encircle it everywhere, covering it with transverse rings of very variable thickness, from its junction with the trunk to its extremity. This latter contracts to an inextensible, and usually short cord or peduncle, which itself terminates in a sort of disk. In the foot are also the two club-shaped glands (so common in other genera) which probably secrete a viscous fluid for fastening the disk to some extraneous object.

The **Nutritive System.**—The food is feebly pecked at by the jaws, while it is in the crop, but it evidently undergoes there some process of digestion. I once saw a Floscule bolt a small *Salpina.* When inside the crop it was still alive, and it charged from side to side, in the vain hope of escape. The sharp points of its lorica ought to have made its captor uncomfortable ; but the only result was, that its outline gradually grew dim, and that before long the whole animal faded into a shapeless mass.

Mr. Gosse[3] notices the absence of the mastax, and says of the **trophi** that "the jaws consist of a pair of curved unjointed but free mallei, with a membranous process beneath each. Each malleus (Pl. I. figs. 9a, 9b) is an uncus of two slender arched divergent fingers, united by a subtle web ; the back of each curves downwards, where, expanding and becoming membranous, it is connected with some delicate but definite processes with rounded outlines, which I should have supposed to be muscular bulbs, but that they remain after treatment with potash."

After passing between the jaws the food enters the stomach (Pl. II. fig. 4, s) apparently directly, as no œsophagus is visible. The alimentary canal is divided into stomach (s) intestine (i), and rectum (r), which latter is bent on itself, and ascends to the cloaca (cl)

---

[1] Fig. 1. Pl. P. is a copy of Dr. Moxon's drawing, showing the tube turned inside out and protruded through the slit in the diaphragm ; the lobes (l) are shown drawn down close to the body.
[2] Popular Sci. Rev. vol. i. 1862, p. 160, pl. ix.　　[3] Phil. Trans. 1856, p. 119.

on the dorsal surface. A partial separation (probably due to a sphincter muscle) is usually visible between the stomach and intestine, and the distinction is often made more obvious by the different colour of the contained food. The whole alimentary tract is richly ciliated; and so is the tube (*t*) that hangs down into the crop. The cilia can be easily seen in the intestine, when it is partially empty; and on the tube, when it is everted by the Floscule's disgorging its crop.

Secreting System.—In the great majority of Rotifera there are two gastric glands, seated on the top of the stomach. I have, however, never been able to detect them in any Floscule, neither has anyone recorded their existence except Ehrenberg and Grenacher. The latter describes and figures them of unusual length in *F. campanulata*,[1] but it is probable that in this matter he is mistaken. He also describes two club-shaped glands in the foot.

Vascular System.—It is very difficult to trace this in most of the Floscules, owing to the optical difficulties due to their skins. But in 1864 Dr. Moxon (*loc. cit.*) published a complete account of it in *F. campanulata*. His figure of the contractile vesicle (Pl. II. fig. 3, *cc*), the lateral canals (*lc*), and of four of the vibratile tags (*vt*), is so clear as to render any verbal description unnecessary. A fifth vibratile tag was discovered by Grenacher in *F. campanulata* (*loc. cit.*), in the side of the coronal cup, near the spot where Dr. Moxon (*loc. cit.*) had anticipated that it would, some day, be found. Parts of this system have been seen in several other species, and doubtless it exists in all.[2]

The **muscles** consist of non-striated fibres. Below the bottom of each depression, between the lobes, a muscle runs downwards in the substance of the coronal cup and vestibule, and is lost on the surface of the body, to reappear again, as it nears and passes down the foot. The anterior portions of these muscles end in two or more branches which diverge to the thickened rim of the coronal cup, and often interlace, as seen in *F. coronetta* (Pl. II. fig. 2), and in *F. trilobata* (Pl. II. fig. 6), where they may be seen reaching the summit of the dorsal lobe. There are some half-dozen transverse muscles imbedded in the integument of the trunk; and the walls of the vestibule, with its upper ciliated rim, are all highly contractile.

The Nervous System.—Dr. Moxon (*loc. cit.*) has described and figured the nervous ganglion in *F. campanulata*, and I have seen it in *F. Hoodii*. It is in nearly the same position as it is in *Stephanoceros*, namely, on the dorsal side of the vestibule; and is, as usual, close to the organ of taste, and not far from the eyes and dorsal antenna; to all of which doubtless it sends out nervous threads. Dr. Moxon has seen and figured such threads in *F. campanulata* (Pl. II. fig. 3, *n.*)

In the great majority of the Rotifera the mastax is also not far from the nervous ganglion; but in the *Flosculariadæ* the mastax almost vanishes, while the jaws and ganglion are far apart.

Organs of Sense.—Two red eye-spots lie usually above the ganglion; but, as in the adults they are deeply imbedded in the integument, they are not easily seen. In fact the ordinary way of attempting to see them, by transmitted light, will scarcely ever be successful; but by treating the Rotiferon as an opaque object, and concentrating a strong light on it, the eyes may often be seen glowing like rubies when all else is invisible.[3] The **eyes** are conspicuous in the half-grown animals, and in the young within

[1] *Sieb. u. Köll. Zeits.* Bd. xix. 1869, p. 483.

[2] Dr. Leydig saw the contractile vesicle in *F. cornuta*; *Ueb. d. Bau d. Räderth.* 1854. Dr. Bartsch has seen the contractile vesicle, lateral canals, and vibratile tags in his *F. longilobata* (*F. coronetta*) *Rot. Hungariæ*, 1877.

[3] Mr. Gosse (*Popular Sci. Rev.* vol. i. 1862, p. 166) observed one eye at a time in *F. cornuta*. Mr. Cubitt (*Mon. Micr. J.* vol. iii. 1870, p. 215) saw the eyes in *F. coronetta*; and I have recorded (*J. Roy. Micr. Soc.* 2 Ser. vol. iii. 1883, p. 163) my having observed them in *F. Hoodii* and *F. regalis.* Herr K. Eckstein (*Sieb. u. Köll. Zeits.* Bd. xxxix. 1883, p. 347), unaware of the above observations, says "the Floscules have been hitherto regarded as eyeless," and records his having seen the eyes in *F. cornuta*.

the egg. There are three **antennæ** in *F. campanulata*, *F. coronetta*, and *F. Hoodii*, and no doubt the same three may with care be found in the other species. There is one on each side of the neck, and one on the median line near the middle of the dorsal lobe. The two lateral antennæ are very short tube-like prominences each carrying a brush of divergent setæ; they are very apparent in *F. coronetta* (Pl. II. fig. 2): the dorsal antenna is a mere setigerous pimple.[1] The **setæ** on the lobes act also as organs of touch, warning the creature of the approach of anything detrimental to its delicate cup; and whipping back into it any animalcule that endeavours to escape from it.

On the middle of the contractile collar, which is above the vestibule, and on the dorsal side of it, there is a round projection facing the concavity of the ciliary wreath. It can be easily seen in *F. coronetta* and *F. Hoodii*, and is probably an organ of taste, as it is constantly thrust forward to meet any particle which is passing into the vestibule. A very obvious and tongue-like organ holds a similar position, and acts in a similar way, in *Stephanoceros*.

The **Reproductive System.** The ovary, with its clear spherical germs, and frequently with an opaque egg in it, can be seen filling the greater part of the space between the stomach and the ventral surface. No other portion of the apparatus has been made out, owing no doubt to a habit that the Floscules have of contracting themselves sharply into their tube when about to lay an egg. When laid, the eggs are ranged above one another, between the foot and the tube. The ordinary number of female eggs is from two to five; though as many as eight or ten have been seen at once. The male eggs, which are smaller, rounder, and more numerous, than the female, frequently amount to as many as nine or ten, and have occasionally been seen in a cluster of eighteen or twenty in the same tube. Both are inclosed in a delicate shell, which is left behind in the tube, when the young Floscule is hatched. Dr. Weisse and Mr. Hood agree in assigning six or seven days as the time from the extrusion of the egg to the birth of the young animal.

The **Young Female.**—"The infant female Floscule is a white cylindrical maggot Pl. I. fig. 9c), blunt at the front end, with a central orifice, whence protrudes a short brush of cilia; but the margins are capable of unfolding, when the cilia are seen to form a whorl around the truncate summit, swiftly rotating. The margin soon begins to bud forth the little knobs around which the cilia are gathered (Pl. I. fig. 9d); these quickly increase in length, and the angular flower-like corona gradually forms. Meanwhile the little creature, which was at first free, attaches itself by its hinder end, and assumes the condition as well as the form of the parent."[2]

Mr. Hood has observed in *F. calva*, that the young animal fixes itself two or three hours after it has burst its shell, and soon begins to form its tube, which at first rises barely to half the height of the foot. By the time it is three days old (Pl. III. fig. 3a) the tube has attained fair proportions.

The same observer noticed that the lobes of the young *F. ambigua* began to develop from a collar under the ciliary wreath, and were at first merely a dorsal and ventral lobe; the latter with a small notch. In three or four days the notch deepened and widened so that there were three lobes; but it was not till the fifth or sixth day that the rudiments of the small side lobes (the fourth and fifth) made their appearance. The young Floscule arrived at maturity at the twenty-fourth or twenty-sixth day, but continued to increase in size after it had deposited eggs; in fact, did not cease to grow till shortly before its death. The whole lifetime, in a trough, was from forty to forty-six days.

Captivity, however, affected the growth of the animals, even when carefully attended to, and plentifully supplied with food. On one occasion, for instance, a large *F. campanulata* ¹⁄₅ inch long, from one of the Scotch lochs, was placed in a tank; and

---

[1] Dr. Moxon (*l'c. cit.*) first called attention to their existence in the Floscules. Herr Grenacher (*loc. cit.*) mentions his having discovered the median antenna in *F. proboscidea* (*F. campanulata*); but was unaware that Dr. Moxon had seen and described all three, five years before.

[2] Mr. Gosse on *F. campanulata*. *Popular Sci. Rev.* vol. i. 1862, p. 166.

the young reared from its eggs, though perfectly healthy and breeding freely, never exceeded ¹⁄₄ inch in length : their eggs, too, were half the size of those of their parent.

In *F. trilobata*, occasionally, the egg produces the living young in the body of the parent. Mr. Hood has seen the embryo alive in the egg, within the Floscule, and has witnessed its birth : yet Dr. Collins has seen the same Rotiferon deposit the usual eggs in its tube.

The **Male.**—Until 1874 no male had been discovered among the *Rhizota* ;[1] and indeed some observers supposed this group to be monœcious ; but in that year I had the good fortune to find the male of *Lacinularia socialis*, and to study it thoroughly.[2] Soon afterwards I found that of *Floscularia campanulata* (Pl. I. fig. 1c) and I have since seen what I believe to be the male of *F. mutabilis* (Pl. III. fig. 2c). Mr. Hood has observed and figured the male of *F. calva* (Pl. III. fig. 3b), and has seen that of *F. ambigua* actually hatched. The structure of the male Floscule has not yet been thoroughly investigated ; but, so far as it has been studied, it has been found to agree with that of other male Rotifera. The corona is an imperforate many-lobed cushion, surrounded by a simple circlet of long cilia. The **nutritive system** is wholly absent. Two red **eyes** are visible just under the surface of the corona ; and the longitudinal **muscles**, for withdrawing the head, are generally obvious. Nearly the whole of the body-cavity is filled with a large **sperm-sac** (Pl. I. fig. 1c, and Pl. III. fig. 3b ; ss) from which the **penis** (p), a ciliated protrusile tube, proceeds to the dorsal surface, at the junction of the trunk and foot.

The **vascular system, ganglion,** and **antennæ** have not yet been seen ; but no doubt they are present, as in the males of other Rotifera.

F. REGALIS, *Hudson.*

(Pl. I. fig. 8.)

Floscularia regalis   Hudson, *J. Roy. Micr. Soc.* 2 Ser. vol. iii. 1883, p. 166, pl. iv. fig. 5.

SP. CH. **Lobes** *seven, knobbed.*

The **corona** is a deep cup with a nearly circular rim, from which project four knobbed triangular processes on the ventral side, dividing that half of the rim into three equal spaces. The processes curve slightly outwards ; and, at the rim, their bases unite, so as to give that edge of the cup a semi-hexagonal appearance. In the middle of the dorsal side of the rim rises a large triangular knobbed lobe, bearing on each side a short recurved knobbed process. All seven knobs carry pencils of long radiating setæ. A true **ciliary wreath** at the bottom of the trochal cup, and two red **eyes**, can be easily seen. This remarkable Rotiferon, the only seven-lobed species, was found by Mr. Thos. Bolton in September 1882, near Birmingham.

Length, ¹⁄₅₀ to ¹⁄₄₀ inch.[3] **Habitat.** Lakes and clear ponds. Birmingham (T.B.[4] ; Perth (J.H.[4] and W. Dingwall) : not common.

F. CORONETTA, *Cubitt.*

(Pl. I. fig. 5 ; Pl. II. fig. 2.)

Floscularia coronetta   Cubitt, *Mon. Micr. J.* vol. ii. 1869, p. 133, pl. xxv.
Stephanoceros Horatii   Cubitt, *Mon. Micr. J.* vol. vi. 1871, p. 166.
Floscularia longilobata   Bartsch, *Rot. Hungariæ,* 1877, p. 21, ii. Tab. ii. 14.

---

[1] Mr. Gosse (*loc. cit.* p. 187) described some probably male eggs in *M. ringens.*

[2] *Mon. Micr. J.* vol. xiii. 1875, p. 15.

[3] As the Rotifera vary from ¹⁄₄ to ¹⁄₁₅ inch, no attempt has been made to draw them to a fixed scale. The actual length of each species will be given at the end of its description.

[4] Throughout the work the following initials will be used in the Habitat : J. H.—Mr. John Hood ; T. B.—Mr. Thos. Bolton ; P. H. G.—Mr. Gosse ; C. T. H.—Dr. Hudson.

SP. CH. **Lobes** *five, linear, knobbed ;* **setæ**. *non-extensile.*

The **corona** has five long narrow knobbed lobes, nearly all of equal length, separated by deep depressions, and forming a miniature coronet. The dorsal lobe is slightly the longest, and the lobes are so set on the front of the body that a plane touching the knobs would be oblique to its longitudinal axis; the dorsal lobe being the furthest forward. All the knobs carry long radiating setæ, and the setæ are continued all along the edge of the trochal cup (Pl. II. fig. 2). The true **ciliary wreath** and the **eyes** have been seen in the adult by Cubitt (*loc. cit.*); but the former with difficulty. The lateral **antennæ** can be readily seen when the animal is favourably placed, as well as the delicate muscular threads by which the longitudinal **muscles** act on the corona (Pl. II. fig. 2). As many as seven **male** eggs have been seen in one tube.

Length. $\frac{1}{24}$ inch. **Habitat.** In ponds and marsh pools ; rare. Wandsworth Common (Cubitt) : Forfar, Fife (J.H.).

F. MIRA, *Hudson.*

(Pl. III. fig. 1.)

Floscularia mira . Hudson, *J. Roy. Micr. Soc.* 2 Ser. vol. v. 1885, p. 609.

SP. CH. **Lobes** *five, linear, knobbed ;* **setæ** *extensile.*

The **corona** is very like that of *F. ornata*, which species the Rotiferon closely resembles in every respect but two. First, the **tube** is much more like that of a *Stephanoceros* than that of an ordinary Floscule. I have seen only one specimen, but Mr. Cocks (its discoverer) tells me that the tubes of the half-dozen specimens which he has seen were all of the same sort. Secondly, in its **setæ** *F. mira* is not only unlike all other Floscules, but is unique among the Rotifera : for each seta is in constant independent motion, slowly extending or contracting like the pseudopodium of an Amœba. When the retracted seta begins to extend again, it is often bent into a whip-like shape, a wave of motion overtaking, as it were, the resting anterior portion, and finally driving out the latter with a characteristic flourish of its tip. The setæ are of amazing length and abundance, exceeding the total length of the Rotiferon. This very rare and wonderful creature was discovered by Mr. W. G. Cocks in June 1884.

Length. $\frac{1}{30}$ inch. **Habitat.** Unknown ; found in an aquarium, in water that probably came from Epping Forest or Walton-on-Thames (W. G. Cocks).

F. ORNATA, *Ehrenberg.*

(Pl. I. fig. 9.)

Floscularia ornata. Ehrenberg, *Die Infus.* 1838, p. 408, Taf. xlvi. fig. 2.
,, ,, Peltier, *Ann. Sci. Nat. Zool.* 2 Sér. t. x. 1838, p. 11, pl. iv.
,, ,, Dujardin, *Hist. Nat. Zooph.* 1841, p. 610, pl. xix. fig. 7.
,, ,, Gosse, *Tenby,* 1856, p. 307, pl. xx.
,, ,, ,, *Popular Sci. Rev.* vol. i. 1862, p. 160, pl. ix. figs. 1-3.
,, ,, Pritchard, *Infusoria,* 1861, p. 675, pl. xxxii. figs. 384, 385, and xl. figs. 25, 26.
,, ,, Weisse, *Sieb. u. Köll. Zeits.* Bd. xiv. 1864. p. 107, Taf. xiv. figs. 1-5.
,, ,, Bartsch, *Die Räderth. b. Tübingen,* 1870, p. 21.

SP. CH. **Lobes** *five. triangular, knobbed ;* **dorsal lobe** *without any process.*

The **corona** has five knobbed lobes of moderate length separated by broad depressions, the dorsal lobe being distinctly the longest and broadest, and the knobs crowned with long radiating setæ. Ehrenberg describes *F. ornata* as having usually six lobes, but sometimes five, and draws an example of each case. No doubt it must have been a difficult matter for one of the old observers with only a monocular microscope, and that a poor one, to make out the shape of a delicately transparent and scolloped cup, presented to him so that its upper and under surfaces were projected on each other. With a modern binocular and dark-field illumination, no tyro would fail to describe correctly the cup with its five knobbed lobes. Ehrenberg credits this species with two "clear spaces" that he considers to be gastric glands. I have made frequent search for such glands, but cannot find them; Ehrenberg's "clear spaces" are probably the small bulbs, the rudiments of a mastax, in which the jaws are inserted.

It is most probably Eichhorn's "Der Fänger" (Pl. B, figs. 15, 16), and, if so, it is the earliest known Floscule, having been discovered as long ago as 1767. Unluckily, Eichhorn has given two other drawings of it, one with nine, and one with ten knobbed lobes ; but, as he complains of the difficulty of rightly understanding and drawing it, it is possible that these numerous lobes represent only the puckers of the half-expanded corona. This is a very pretty species, and, as Eichhorn well says, "no lightning can dart out of the clouds through the air more swiftly" than this little animal can contract upon its prey. Owing to its small size, however, and its lack of transparency, it is not well adapted for the investigation of the internal organs of the Floscules.

Length. From $\frac{1}{60}$ to $\frac{1}{45}$ inch ; average $\frac{1}{50}$. **Habitat.** Fresh waters everywhere

F. CORNUTA, *Dobie*.

(Pl. I. fig. 7.)

Floscularia cornuta	. .	Dobie, *Ann. Nat. Hist.* 2 Ser. vol. iv. 1849, p. 233, with pl.
" "	. .	Gosse, *Popular Sci. Rev.* vol. i. 1862, p. 168, pl. ix. fig. 6.
" "	. .	d'Udekem, *L'Institut*, t. xix. 1851, p. 222.
Floscularia appendiculata	.	Leydig, *Ueb. d. Bau d. Räderth.* 1854, p. 3, Taf. i. fig. 6.
Floscularia cornuta	.	Pritchard, *Infusoria*, 1861, p. 676, with fig.
" "	.	Cubitt, *Mon. Micr. J.* vol. v. 1871, p. 170, pl. lxxxi. fig. 7.
Floscularia appendiculata	.	Bartsch, *Rot. Hungariæ*, 1877, p. 24, Táb. ii. fig. 19.
" "	.	Eckstein, *Sieb. u. Köll. Zeits.* Bd. xxxix. 1883, p. 344, Taf. xxiii. figs. 1-4.

SP. CII. **Lobes** *five, triangular knobbed ;* **dorsal lobe** *with flexible process.*

This species was first described by Dr. Dobie (*loc. cit.*) and was afterwards re-named as a new species by Dr. Leydig (*loc. cit.*). It is like *F. ornata*, but possesses at the back of the dorsal lobe a curious flexible **process**, which is probably an organ of touch, though it does not appear to bear seta, or to have any opening. It rises from a swollen base just below the knob, and is suddenly bent over the latter, and then turned up again so as to point forwards and clear the knob. It occasionally moves a little, and slowly alters its shape, taking often an undulating form ; but it is not moved about like the antenna of *Cephalosiphon* or of *Rotifer macroceros* : it reminds one rather of the slow bendings of the dorsal appendages of *F. Hoodii*. The **eyes** cannot be easily seen in the adult, but I have succeeded in exhibiting both together by condensing a strong lamplight on the dorsal surface.

Leydig (*loc. cit.*) describes and figures the **contractile vesicle** but places it away from the intestine on the ventral side.

Length, cir. $\frac{1}{50}$ inch ; Scotch specimens up to $\frac{1}{40}$ inch. **Habitat.** Widely distributed.

F. CYCLOPS, *Cubitt.*

(Pl. I. fig. 6, and Pl. D. fig. 2.)

Floscularia cyclops	Cubitt, *Mon. Micr. J.* vol. vi. 1871, p. 83, pl. xciii. figs. 1, 3.

SP. CII. **Lobes** *five, knobbed, very short, variable in length, but sometimes with the knobs almost seated on the rim of the coronal cup ;* **the dorsal lobe** *rather the longest and stoutest ;* **setæ** *radiating from the knobs.*

E 2

This Floscule greatly resembles *F. ornata*, but it is distinguished by its height,[1] the length of its foot, and the shortness of its lobes. The fully extended **foot** is frequently thrice as long as the body. The **tube** is much wider than usual in proportion to the animal's size, and often symmetrical in shape, like that of *F. longicaudata*. Two **eyes** are visible in the adult. This species is prolific and has often many eggs in its tube. As many as twelve female eggs have been counted in the same tube; and eighteen male eggs in another. Found by Mr. C. Cubitt in 1871.

Length, $\frac{1}{35}$ inch. **Habitat**. North Brook, Kent (Cubitt); ponds and marsh pools, Forfar, Fife (J.H.); rare.

F. CAMPANULATA, *Dobie*.

(Pl. I. Fig. 1.)

Floscularia proboscidea .	Ehrenberg, *Die Infus.* 1838, p. 408, Taf. xlvi. fig. 1.
,, ,,	Dujardin, *Hist. Nat. Zooph.* 1841, p. 610.
Floscularia campanulata	Dobie, *Ann. Nat. Hist.* 2 Ser. vol. iv. 1849, p. 233, with pl.
,, ,, .	Gosse, *Popular Sci. Rev.* vol. i. 1862, p. 167, pl. ix. figs. 1, 5.
,, ,, .	Pritchard, *Infusoria*, 1861, p. 675, with fig.
,, ,, .	Hudson, *Trans. Bristol Micr. Soc.* 1867, 2 pls.
Floscularia proboscidea	Grenacher, *Sieb. u. Köll. Zeits.* Bd. xix. 1869, p. 483, with fig.
Floscularia campanulata	Cubitt, *Mon. Micr. J.* vol. viii. 1872, p. 5, pl. xxiv. fig. 1.

SP. CII. Lobes *five, broad, without knobs, separated by distinct depressions:* **peduncle** *short;* **setæ** *radiating from the summits of the lobes, and fringing the whole edge of the coronal cup.*

I think that Grenacher (*loc. cit.*) is right in supposing that Dr. Dobie's *F. campanulata* is really Ehrenberg's *F. proboscidea*. Ehrenberg describes the latter as having six lobes, and also a snout-like organ, of cylindrical form, beset with setæ like those on the lobes, and rising from the depths of the coronal cup above its rim. Grenacher suggests, as Dujardin had done before him, that this snout-like organ is only the dorsal lobe seen before the corona is fully expanded. I have thought it best, however, to retain Dr. Dobie's name, as *F. campanulata* has certainly neither a proboscis nor six lobes: I confess, however, that I have little expectation of anyone's ever finding a Floscule with either the one or the other.

The **setæ** often appear to be confined to the thickened summits of the lobes, forming simply a tuft on each. They really, however, fringe the whole circumference of the corona, sloping further away from it as they approach the bottoms of the depressions between the lobes, and even at last pointing backwards towards the foot. The **vascular system** has been described above, p. 47, and is shown Pl. II. fig. 3. Only four vibratile tags are given in Dr. Moxon's figure; but Grenacher has seen a fifth, whose position is shown in Pl. II. fig. 4. A **nervous ganglion** has been seen by Dr. Moxon. It is situated dorsally on the neck (Pl. II. figs. 3 and 4, *gn*). Nerve threads are drawn by Dr. Moxon, as passing from the ganglion to the three antennæ. There is one dorsal antenna, half-way up the coronal cup, and one on each side of the cup close to its junction with the body (Pl. II. fig. 3, *a*). They are little more than setigerous pimples. When the coronal cup is furled, the dorsal antenna may be seen on the summit of the contracted Floscule's pear-shaped body. The discovery of the **male** has been mentioned above, p. 19. Its sperm-sac (*s*) and penis (*p*) are indistinctly shown in Pl. I. fig. 1, *c*; but the dead specimen from which I drew the figure was so lately hatched that its cuticle was more than usually opaque. As many as twenty male eggs have been

---

[1] To obtain a correct notion of the completely expanded animal, the foot and case in Pl. I. fig. 6, should be supposed to be continued quite two inches below the bottom edge of the page on which the figure is drawn. A small, correctly proportioned figure, is given in Pl. I. fig. 2.

counted [1] in one tube. This beautiful Rotiferon is by no means shy, but often attaches itself in closely-packed clusters, of a dozen or more, to the stems or ends of the leaves of water-plants.

Length. Average about $\frac{1}{12}$ inch ; but Mr. Hood has found in the Scotch lochs specimens no less than $\frac{1}{10}$ inch long. **Habitat.** Clear ponds and lakes : common.

F. LONGICAUDATA, *Hudson*.

(Pl. I. fig. 4.)

Floscularia longicaudata . . Hudson, *J. Roy. Micr. Soc.* 2 Ser. vol. iii. 1883, p. 165, pl. iv. fig. 2.

SP. CII. **Lobes** *five, rather pointed, without knobs, the dorsal lobe the largest, the two ventral ones next in size, and the two lateral ones much the smallest;* **peduncle** *very long;* **setæ** *as in F. campanulata.*

This Floscule resembles both *F.* campanulata and *F.* ambigua but is distinguished from them by its pointed lobes, very long peduncle, and comparative smallness of its body. The lateral **lobes** vary in size in different specimens, and even in the same animal at different times, and are occasionally as minute as they always are in *F. ambigua.* The **peduncle** (*pd*) is often $\frac{1}{3}$rd of the length of the extended foot, while in other Floscules it varies from $\frac{1}{5}$th to $\frac{1}{20}$th of that length. It is a thin, transparent, non-retractile thread, and is generally thrown into graceful curves and coils. The **tubes** of all the specimens which I have seen were remarkably compact and symmetrical.

This is a social Rotiferon, and is to be found sometimes in clusters of a dozen or more, of various ages and sizes. It selects exposed situations, perching itself on the edge or point of a leaf, and preferring the convex side to the concave. It is a great feeder, swallowing small live *Infusoria* greedily ; and, though not so hardy as *F. ambigua,* yet it will bear being kept in a trough for a fortnight.

F. longicaudata was first discovered by Mr. J. Hood in 1881, on a leaf of *Sphagnum* in a pool on Tent's Muir ; and again in Loch Rea, Blairgowrie, in July and August of the same year.

Length. From $\frac{1}{15}$ to $\frac{1}{12}$ inch. **Habitat.** Lochs and marsh pools ; Forfar, Fife, Perth (J.H.) : rare.

F. AMBIGUA, *Hudson*.

(Pl. I. fig. 2.)

Floscularia ambigua . . Hudson, *J. Roy. Micr. Soc.* 2 Ser. vol. iii. 1883, p. 163, pl. iv. fig. 1.

SP. CII. **Lobes** *apparently three ; viz. one large broad dorsal lobe, and two much smaller ventral ones ; a pair of minute lateral lobes lie between the dorsal and ventral lobes ;* **setæ***, as in F. campanulata.*

This broad stumpy Rotiferon connects the five-lobed with the three-lobed Floscules ; for though, at first sight, it seems to have but three lobes, there is also a minute lateral pair. These lateral lobes are frequently reduced to mere thickenings of the rim of the cup, but can always be detected by the setæ radiating from them. From some points of view *F.* ambigua closely resembles *F.* campanulata ; and, indeed, I think that Dr. Moxon (Pl. II. fig. 3) may have mistaken the one for the other.

From the body up to the dorsal lobe, as in *F. Hoodii,* run two ridges of semi-transparent tissue, which look like buttresses to the coronal cup ; and form, with it and the dorsal surface, a deep hollow, at the bottom of which lies the neck.

The animal has a habit of so contracting itself as to throw its cuticle into deep folds, especially at the neck, and at the base of the body. There often appears also to be a

[1] By Mr. W. Dingwall, of Dundee.

well-marked separation between the body and foot, the latter looking as if it possessed only half the width of the body, at the line of junction.

F. ambigua was discovered by Mr. J. Hood in May 1881 on a leaf of *Sphagnum*, in a mossy pool on Tent's Muir, Fife. Its habits are the reverse of those of *F. longicaudata*. It selects for its post the axil of a plant, or the under surface of a leaf, especially of a well-curled one ; so that it is difficult to find a specimen that can be easily studied from various points of view. Thus placed as it were in ambush, the burly Floscule draws, with its powerful ciliary wreath, all kinds of organisms into its coronal cup. Nothing seems to come amiss to it, and its appetite never fails. Mr Hood has seen it devour the young of *Œcistes pilula*, and of *Œ. umbella* ; as well as other free swimming Rotifera, along with all kinds of *Infusoria* ; so that, to use his own vigorous language, "it would eat its own weight in three hours." The same observer has twice seen the **male** hatched from the egg laid in the tube ; and noticed the motion of its spermatozoa in the sperm-sac.

Length. From $\frac{1}{50}$ to $\frac{1}{40}$ inch. **Habitat.** Lochs and marsh pools ; Forfar, Fife, Perth (J.H.); near Birmingham (T.B.) ; Woolston pond (P.H.G.) : sometimes abundant.

F. ALGICOLA, *Hudson*, sp. nov.

(Pl. I. fig. 3 : Pl. II. fig. 1.)

SP. CH. *Very small ; corona precisely that of F. ambigua, but ornamented with dots arranged in symmetrical patterns ; tube, if present, undistinguishable.*

This pretty little Rotiferon is very like *F. ambigua*, differing but little from it except in its ornamented corona, small size, and strange dwelling-place. Its coronal cup is ornamented on the outside with minute dots, arranged in a symmetrical pattern, as shown in Pl. II. figs. 1a, 1b. It makes its home in a parasitic growth (*Gloiotrichia pisum* [1]) on the stems of water plants. Possibly it may in this way avoid the necessity of making a tube, as the parasitical sphere that it lives in seems to consist chiefly of a kind of grey mucus ; but I could not be certain whether it had a tube or not : Mr. Gosse searched with great care, but could see none.

This Rotiferon was found first by Mr. J. Hood in 1882. at Rosemont Loch, Blairgowrie. It was then very abundant.

Length. $\frac{1}{90}$ inch. **Habitat.** Lochs, Perth (J.H.) : not common.

F. TRILOBATA, *Collins*.

(Pl. II. fig. 6.)

| *Floscularia trilobata* | . | . | Collins, *Science Gossip*, Jan. 1872, p. 9, with fig. |
| *Floscularia trifolium* | . | . | Hudson, *J. Roy. Micr. Soc.* 2 Ser. vol. i. 1881, p. 4, pl. ii. |

SP. CH. *Lobes three, large. broadly curved, separated by very deep and similarly curved depressions : dorsal lobe rather the largest ; setæ forming a continuous double fringe round the entire circumference of the corona ; the outer row arranged like those of F. campanulata ; the inner row short, slightly curved, and arranged like cilia.*

This large and elegant Floscule was discovered by Dr. F. Collins in 1865, in a small pool near Sandhurst, Berks ; and he published a short account of it, with a figure (*loc. cit.*) in 1872. It was afterwards found by Mr. J. Hood, in Loch Laudie, near Dundee, in 1880 ; and I published a description of it (*loc. cit.*) naming it *F. trifolium*, as I had considerable doubt of its really being Dr. Collins' species. I have since seen the description and figures which Dr. Collins sent, with some live specimens, to Mr. Gosse in 1865 ; and I have now no doubt that *F. trifolium* and *F. trilobata* are the same.

[1] Kindly identified by Dr. Cooke in a letter to Mr. Gosse.

The peculiarity of a second fringe of setæ lying within the principal fringe is shared with it by *F. Hoodii* alone. Unlike the ordinary setæ, these smaller secondary setæ possess a kind of joint action; for when any captive creature tries to escape from the coronal cup, and to pass the setæ, those of the outer row either lash separately at it, or are drawn together over it by the converging lobes; while a wave of motion, like a ciliary wave, runs once or twice round the inner row. The true ciliary **wreath** can be easily seen at the bottom of the coronal cup, owing to this Floscule's great size and transparency. I saw a small **contractile vesicle**, but I had no opportunity of tracing the rest of the **vascular system**. The two lateral **antennæ** were also obvious.

The first thing that strikes the observer, on watching the protrusion of the furled head, is the great size of the Floscule, and the curiously shrivelled appearance that the lobes of the coronal cup present, as they emerge from the opening head. They look exactly as if the animal were sickly or injured. In a few seconds, however, they gently swell out, the many folds and creases disappear, till at last the eye is gratified with the sight of a lovely transparent tulip, of three petals, their edges all fringed with delicate and motionless hairs. It is a creature of exquisite beauty; from every point of view the flowing curves of the cup are charming, and its great transparency permits the whole of the outline of the rim to be seen at once. The dorsal lobe is rather larger than the other two, and is curved forward over the cup. Across each lobe run delicate muscular threads for furling it, which are specially visible on the back of the dorsal lobe (Pl. II. fig. 6).

Dr. Collins saw **eggs**, laid by his specimens, remaining attached to them within the tube; but Mr. Hood observed that some specimens, which he reared in a trough, hatched the ova in the ovary, and then gave birth to the living young.

They are voracious feeders on *Infusoria* and small animalcules, and are fond of stationing themselves in the axils of water plants.

Length, ¹⁄₁ to ¹⁄₂ inch. **Habitat.** Lochs, marsh and boggy pools; Sandhurst (Dr. Collins and P.H.G.); Woolston, Hants (P.H.G.); Fife, Forfar, Perth (J.H.): not common in England, abundant in Perthshire.

F. Hoodii, *Hudson.*

(Pl. II. fig. 5.)

Floscularia Hoodii . Hudson, *J. Roy. Micr. Soc.* 2 Ser. vol. iii. 1883, p. 161, pl. iii. figs. 1, 2.

SP. CII. **Lobes** *three; dorsal lobe much the largest, and carrying two, large, sleeve-like, flexible processes; setæ, short, cilia-like, in two parallel rows, fringing the entire circumference of the coronal cup.*

This is one of the largest of all the Rotifera; adult specimens being quite ¹⁄₁₀ inch from the top of the dorsal lobe to the extremity of the peduncle. Its great size, and its curiously shaped three-lobed corona make it sufficiently remarkable; but, in addition to these peculiarities, it has two extraordinary processes, perched one on each side of the back of the dorsal lobe. They appear to be hollow, and to communicate with two lenticular spaces lying between the two surfaces of the dorsal lobe. Fine muscular threads pass along and across them (Pl. II. 5b), and the animal can contract and expand each independently of the other; and throw them into all kinds of positions. The upper end of each seems to be separated from the lower portion by a constriction, from which a muscular thread runs down to the base. Each of these processes slowly and independently changes its shape and position, now sinking down on the dorsal lobe so as to be invisible, or again bending its free end at right-angles to the lower portion. I have never seen anything like them on any other Rotiferon: they may possibly be organs of touch, but I could detect no trace of setæ on them. Mr. Hood tells me that both in young and adult specimens he has seen brown granular matter discharged from their free ends. The thickened rim of the three lobes carries its double fringe of setæ set just as they are in *F. trilobata*, the larger row stretching outwards, and the smaller inwards; and the same rapid flicker may be seen on occasions to run all round the edge of the coronal cup. The orifice of the coronal cup alters constantly, now opening in the characteristic way shown in Pl. II. fig. 5a, and now reduced to a slit; or even closed in puckers. Two dorsal ridges, as in *F. ambigua*, run like buttresses from the body to the back of the dorsal lobe, and in the lowest portion of the deep hollow between these lie the two pale pink **eyes**; both in the neck, and one close to each buttress. The true **ciliary wreath** is distinctly visible throughout its whole length. It is a long horse-shoe-shaped and ciliated ridge, sloping sharply down from the bottom of the coronal cup into the vestibule. The **contractile vesicle** is unusually large and distinct; close to it, and apparently situated in it, is a cluster of yellow globules, which look black by transmitted light.

This strange and beautiful Floscule was discovered by Mr. J. Hood in December 1882, in a ditch on Tent's Muir, Fifeshire.

Length, ¹⁄₁₀ inch. **Habitat.** Marsh pools; Fife (J.H.): rare.

F. CALVA, *Hudson*.

(Pl. III. fig. 3.)

Floscularia calva . . Hudson, J. Roy. Micr. Soc. 2 Ser. vol. v. 1885, p. 610.

SP. CH. **Lobes** *two, short*; **dorsal lobe** *the larger*; **setæ** *very short, radiating from the thickened summits of the lobes, incapable of cilia-like action*; **body** *unusually long and narrow, its outline confluent with that of the coronal cup, so that there is no neck*; **eyes** *cervical*.

Mr. J. Hood discovered this species in 1884 on a *Sphagnum* leaf, in a mossy pool on Tent's Muir, only twelve inches deep, and on another occasion found it in Loch Landie at a depth of ten feet. I have seen only two specimens of it, and those under disadvantageous circumstances; as each had dropped from the plant on which it was found, and was lying in the sediment at the bottom of the tube. The creature appears to attach itself rather to its tube than to the stem of the plant which bears the tube, and so to be easily detached. I am indebted to Mr. Hood for drawings of the young male and female (Pl. III. figs. 3a, 3b), each of which he saw hatched from eggs laid in the tube. The male is about ₂¹⁄₁₀ inch in length, and resembles that of *F. campanulata*.

Length, ₂³⁄₁₀ to ₃⁄₁₀ inch. **Habitat.** Lochs and marsh pools, on *Myriophyllum* and *Sphagnum*; Forfar, Fife (J.H.): rare.

F. MUTABILIS, *Bolton*.

(Pl. III. fig. 2.)

Floscularia mutabilis . . Hudson, J. Roy. Micr. Soc. 2 Ser. vol. v. 1885, p. 609, pl. xii. figs. 1 3.

SP. CH. **Lobes** *two, well developed*; **dorsal lobe** *decidedly the larger*; **setæ** *rather short, set round the whole circumference of the disk, and capable of cilia-like motion*; **eyes** *near the summit of the dorsal lobe*.

F. mutabilis somewhat resembles *F. calva*, but is at once distinguished by its larger lobes, moveable setæ, and by its unique habit of swimming. The animal has not as yet been found attached to any water plant. It looks, when resting in its case at the bottom of a live cell, just like an ordinary Floscule that had been knocked off its perch, as the setæ are straight and motionless. After a short rest it pulls down the two lobes to a level with the bottom of the depressions between them, and so alters the **corona** that it looks like that of an *Œcistes*; at the same instant the setæ[1] set up a vigorous cilia-like action; and the animal, ease and all, sails slowly, stern foremost, through the water. Two red **eyes** are very conspicuous in a most unusual position; namely, near the top of the dorsal lobe. I have seen what I believe to to be the **male** (Pl. III. fig. 2*o*), but I failed to isolate it so as to make out its internal organs. Its length was about ₁⅒ inch. It appeared to have, in addition to the usual ciliary wreath, setæ pointing backwards to the foot.

Mr. T. Bolton discovered *F. mutabilis* in a pond of Sutton Park, near Birmingham, in May 1884. He described, named and figured it, soon afterwards, in one of the fly-leaves sent out with his specimen tubes.

Length. About ₂⅒ inch. **Habitat.** A pond in Sutton Park, Birmingham (T.B.) : rare.

F. EDENTATA, *Collins.*

(Pl. III. fig. 1.)

Floscularia edentata	. .	Collins, *Science Gossip,* Jan. 1872, p. 9, with fig.
,, ,,	. .	Hudson, *J. Roy. Micr. Soc.* 2 Ser. vol. v. 1885, p. 611.

Corona *lobeless, transversely truncate;* setæ *very short, chiefly on the central and dorsal portions of the rim;* body *large and stout in proportion to the animal's total length, and nearly as long as the foot.*

Dr. Collins first discovered this ugly Floscule near Sandhurst in 1867. He says (*loc. cit.*) that it has no masticating organs, and that the food passes directly into a capacious stomach. As his specimen was a female (for it laid an egg while in captivity) this is very unlikely. My specimens were so gorged with food that no internal organs could be seen, except the stomach and a portion of the ovary. One of them was literally crammed full of specimens of *Cocconema,* which not only distended the real stomach and the crop, but even protruded above the rim of the coronal cup. How the animal contrived, with its feeble cilia, so to pack itself with these unmanageable diatoms, I cannot imagine.

Length. My specimens, ₁⁄₁₅ inch; Dr. Collins', ₁⅒ inch. **Habitat.** Sandhurst, Berks (Dr. Collins); Woolston, Hants (P.H.G.); Blair Athol (W. Dingwall) : rare.

Genus ACYCLUS, *Leidy.*

GEN. CII. *One dorsal, frontal* lobe; setæ *absent, the coronal cup edged with a delicate festooned membrane; termination of* foot *truncate.*

* ACYCLUS INQUIETUS. *Leidy.*[2]

(Pl. D. fig. 3.)

Acyclus inquietus	. .	Leidy, *Proc. Acad. Nat. Sci. Pa.* 1882, p. 243, pl. ii. figs. 1-6.

The structure of this species has been only imperfectly made out; but, so far as it has been, the animal appears to be closely allied to the *Floscularia*; and so also do the next two species, *Apsilus lentiformis,* and *Apsilus bipera.* The characters of *Acyclus inquietus* given by Prof. Leidy are as follows :—

" *Body fusiform, tapering behind into a long narrow tail-like appendage, by which it is attached, not distinctly annulated, but becoming transversely wrinkled in con-*

---

[1] It is possible that there may be (as Mr. Bolton says) a row of short cilia round the coronal cup, as well as the larger setæ; but my impression is that there is not: I altered my opinion more than once, while watching the creature, but came at last to the conclusion that it swam by means of its setæ, and not by a subsidiary row of cilia.

[2] Throughout the work the species which are not known to be British will be marked with an asterisk.

traction. A non-ciliated cup-like head prolonged into an incurved digitiform appendage (as a substitute for the usual trochal disc), contractile and retractile."

The Professor found eight specimens of this strange creature, each surrounded by a group of *Megalotrocha alboflavicans*, and all attached to the tubes of *Plumatella diffusa*, in the Schuylkill river, U. S. It is considerably larger than *M. alboflavicans*, and can be readily distinguished with the naked eye, towering above the surrounding cluster of *Megalotrochæ* "like a giant in a crowd." It is a very difficult animal to observe, as it bends abruptly in different directions; suddenly contracting and slowly elongating, and scarcely ever for a moment remaining erect. It is translucent, whitish, with the thicker portion of the body of a yellow or brown hue, due to the colour of the alimentary canal. The **corona** is a cup prolonged on the dorsal side into an incurved lobe (Pl. D, fig. 3). It is capable of being expanded or contracted, protruded or retracted; and when expanded, the dorsal lobe is also extended, but remains somewhat incurved. There are no **cilia** or **setæ** on the edge of the cup or lobe, but both of them are bordered by a delicate festooned membrane. When contracted, the lobe is rolled up spirally (Pl. D, fig. 3b). A narrow, transversely wrinkled neck lies between the cup and the body. No ciliary wreath has been noticed within the cup. There is generally no **tube** present; but in two instances the animal has been seen in a " copious colourless gelatinous sheath." The cup converges into a pouch (the **vestibule**) occupying the neck, which is seen to expand and contract from time to time. Longitudinal **muscles** extend from the neck to the membrane surrounding the coronal cup, passing along its walls. Retractor muscles stretch from the body down the length of the foot. The **secreting, vascular,** and **nervous systems** have not been observed; neither have any **eyes** or **antennæ** in the adult female. The **ovary** is in the usual ventral position, and the ova are large, and unsegmented when extruded.

Length, $\frac{1}{24}$ to $\frac{1}{7}$ inch. **Habitat.** Schuylkill river, U. S. (Prof. Leidy): rare.

Genus APSILUS, *Metschnikoff.*

GEN. CII. Coronal cup *wholly membranous;* **setæ** *and* **foot** *absent.*

* APSILUS LENTIFORMIS, *Metschnikoff.*

(Pl. D, fig. 4.)

Dictyophora vorax (?)	Leidy, *Proc. Acad. Nat. Sci. Pa.* 1857, p. 204, and 1882, p. 218, pl. ii. fig. 7.
Apsilus lentiformis .	Metschnikoff, *Sieb. u. Köll. Zeits.* Bd. xvi. 1866, p. 346, with figs.
Cupelopagus lacinedax (?) .	Forbes, *Amer. Mon. Micr. J.* 1882, pp. 102, 151, with fig.
Apsilus vorax . . .	Foulke, *Proc. Acad. Nat. Sci. Pa.* 1884, p. 37 pl. i. figs. 2, 5.
Apsilus lentiformis . . .	Leidy, *Proc. Acad. Nat. Sci. Pa.* 1884, p. 50.

Herr E. Metschnikoff found many specimens of this Rotiferon at Giessen in 1866, on the under side of the leaves of the yellow water-lily, to which they were attached by a chitinous ring on the ventral surface; both in the young and adult female the foot was absent. The **coronal cup** is wholly membranous, and destitute of either cilia or setæ. It is capable of having its edge all drawn close together into a point (Pl. D, fig. 4a), and of being wholly withdrawn within the body, so that it acts as a net, closing over any prey that voluntarily enters it, and forcing it down into the chamber below it, which in the Floscules would be called the **crop** (fig. 4a, cp). At the bottom of the crop is a very peculiar set of **trophi** (fig. 4c). A broad **stomach** has a cæcal appendage on each side, and a cloacal orifice on what appears to be the ventral side, but is really a portion of the dorsal, having been drawn round by the animal's curving its body when attached to the leaf. There are two pear-shaped **glands** attached by their narrow ends to the crop. There is a contractile **vesicle** opening into the cloaca, and from it, above, issues a duct which divides into two lateral canals. Each canal runs slantingly up to a coil at the side of the body below the cup, and thence sends a branch into the dorsal surface of the cup, anastomosing with its fellow above and below the nervous ganglion (Pl. D, fig. 1*b*), and bearing two **vibratile tags** on each side of it. The **nervous ganglion** is a four-side organ in the dorsal wall of the cup; it sends out a nerve thread at each corner, the lower pair passing to two lateral **antennæ**. No **eyes** are visible in the adult. The young **embryo** is developed in the egg in the body of the parent. When hatched, it is a free-swimming Rotiferon (Pl. D, fig. 4*d*), with a truncate, ciliated, anterior extremity, and with the cloaca, at the ciliated posterior extremity, surrounded by a membranous ring. There are two red **eyes**, but the characteristic coronal cup is as yet undeveloped. The **male**, which has a ciliated foot, is so like those already described, that Herr Metschnikoff's drawings supersede description (Pl. D, fig. 4*e*).

Prof. Leidy described in 1857 (*loc. cit.*) a new Rotiferon, "destitute of wheel-organs," which he named *Dictyophora vorax*. He obtained, however, some fresh specimens in 1884, and is now of opinion (*loc. cit.*) that the animal is identical with *Apsilus lentiformis*, and that the discrepancies between his account and Herr Metschnikoff's are due to the wrinkled condition of his first specimens, which had been forcibly removed from the glass sides of an aquarium.

Mr. S. A. Forbes also described (*loc. cit.*) a Rotiferon found in a neglected aquarium, and "wholly destitute of cilia or other vibratile structure." He called it *Cupelopagus bucinedar*, and gave a very characteristic figure of its side view. I have little doubt that this also is *Apsilus lentiformis*.

Length. Maximum about ¹⁄₃₀ inch. **Habitat.** On water plants, Giessen (Metschnikoff); Fairmount Park, and Schuylkill river, U. S. (Leidy).

* APSILUS BIPERA, *Foulke.*

(Pl. D, fig. 5.)

Apsilus bipera   Foulke, *Proc. Acad. Nat. Sci. Pa.* 1881, pp. 37, 50, pl. i. figs. 4, 7.

Apsilus lentiformis .   Leidy, *Proc. Acad. Nat. Sci. Pa.* 1881, p. 50.

Miss S. G. Foulke, who discovered this Rotiferon, is of opinion that it differs from *Apsilus lentiformis* sufficiently to warrant its being regarded as a distinct species ; the points of difference being the shape of the cup, the absence of ganglion, the presence of a " second stomach," and the ciliation of the cup.

If *A. bipera* really has two stomachs, one above the jaws and the other below them, and each a closed sac with walls distinct from those of the body-cavity, then it would not only be a new species, but also a perfectly unique one among the Rotifera. It is evident that further investigation is wanted on this and other points; especially as Prof. Leidy is of opinion that *Apsilus bipera, Dictyophora vorax,* and *Apsilus lentiformis* are all the same animal.

But whether Miss Foulke's species be a new one or not, to her is due the discovery of a true **ciliary wreath** within the coronal cup. It consists of two gradually narrowing ridges, fringed with long cilia, and running up the inside of the dorsal surface of the cup (fig. 5*a*). Short diagonal lines of finer cilia can be indistinctly seen between the larger set. This ciliary apparatus is quite unique in position ; and, if *A. bipera* and *A. lentiformis* are the same, it is curious that Miss Foulke should have missed the nervous ganglion, and that Herr Metschnikoff should have missed the ciliary ridges.

Length. Up to ¹⁄₃₀ inch. **Habitat.** Water-plants in Fairmount Park, U.S. (Miss Foulke).

Genus STEPHANOCEROS, *Ehrenberg.*

GEN. CH. **Lobes** *long, slender, erect, convergent;* **setæ** *set diagonally on the lobes in parallel bands;* **foot** *terminated by an adhesive cup.*

S. EICHHORNII, *Ehrenberg.*

(Pl. IV. fig. 1.)

Stephanoceros Eichhornii	Ehrenberg, *Die Infus.* 1838, p. 400, Taf. xlv. fig. 2.
,, ,,	Dujardin, *Hist. Nat. Zooph.* 1841, p. 612, pl. xix. fig. 8.
,, ,,	Gosse, *Popular Sci. Rev.* vol. i. 1862, p. 30, pl. iii. and iv.
Stephanoceros glacialis	Perty, *Zur Kenntniss kleinst. Lebensf.* 1852, p. 47, Tab. i. fig. 1.
Stephanoceros Eichhornii	Leydig. *Ueb. d. Bau d. Räderth.* 1851, p. 5, Taf. i. figs. 1 1.
,, ,,	Pritchard, *Infusoria*, 1861, p. 668, pl. xxxii. fig. 383, pl. xxxvii. figs. 1 1.
,, ,,	Cubitt, *Mon. Micr. J.* vol. iii. 1870, p. 240, pl. lii.
,, ,,	Newlin Peirce, *Proc. Acad. Nat. Sci. Pa.* 1875, p. 121.
,, ,,	Rousselet, *J. Roy. Micr. Soc.* 2 Ser. vol. iv. 1884, p. 169, pl. v. figs. 1-3.

Anyone who has seen *Stephanoceros* favourably placed, and properly lighted, can well understand the enthusiasm with which Eichhorn relates its discovery[1]; for it is a lovely creature, and as strange as it is beautiful. A small pear-shaped body, whose rich green and brown hues glow beneath a glistening surface, is lightly perched on a tapering stalk, and crowned with a diadem of the daintiest plumes: while the whole is set in a clouded crystal vase of quaint shape and delicate texture. The **tube** is denser than it is in the Floscules, is more symmetrical in shape, and is continuous in substance from its outer surface almost to the creature's body. If an empty tube be examined, it will be found that it has a central hollow, which the body and foot will exactly fill.

Mr. Gosse and Dr. Mantell have each seen a young *Stephanoceros* bore its way through its parent's tube by means of its cilia; just as I have several times seen young Floscules do. The material, therefore, of which it is composed, must be of the flimsiest kind. The commencement of the formation of the tube has been described by Mr. Gosse (*loc. cit.*) as follows: " A specimen, which was hatched under my eye, swam for ten minutes, and then became permanently attached to the upper glass of the box, so that it was vertical in its position, with the foot next to the eye; a favourable aspect for observing the development of the case. It presently began to dilate its body; and, in about five minutes from its attachment, I perceived a distinct filmy ring around it, perfectly circular, whose diameter was about twice that of the body (Pl. IV. fig. 8). The little animal now began to lean over to one side, and the ring soon had another segment additional, leaning in the same direction (fig. 9). The case, for such it was, looked like two broad hoops of glass, each swollen in the middle and set one on the other but not quite concentrically, at least to the eye of the observer. It was manifest that it was produced from an excretion from the body, owing its form and size to the animal's moving round on the foot as on a pivot."

Ehrenberg's drawing of *Stephanoceros* has certainly been taken from a crushed or sickly specimen, and, indeed, in the majority of cases its portrait has been drawn too long after the creature had left its native haunts; for when freshly caught and in vigorous health it arches its five plumes so that its crown almost forms a sphere.[2] The

---

[1] P. 18, *supra.*

[2] Mr. Gosse has found that healthy specimens, removed from an aquarium and inspected at once, have their five arms more frequently produced into a cylindrical form, with their extremities incurved, than arched into a sphere.

setæ are far longer than they appear at first sight, and are not stiff bristles as Ehrenberg has drawn them, but are gracefully curved, and taper off into lines of exquisite fineness. Those of one arm interlace with those of the arms on either side of it, so as to form a living cage of the finest network, through which it is hardly possible for anything to pass without striking some part of the sensitive meshes. The instant this happens band after band of the setæ lashes at the runaway, a swift wave of motion runs along each band, and the captive is thrown back into the vortex produced by the wreath at the bottom of the coronal cup, the ciliary armature of which is precisely like that already described in *Floscularia*. I have also on more than one occasion detected a fitful ciliary wave running round the top of the coronal cup, just under the level of the lowest points of the depressions between its lobes. This has not hitherto been noticed, but I am certain of the fact: the motion was of the briefest duration.

There are considerable differences of opinion about the **muscular system**. Dr. Leydig (*loc. cit.*) says that there are four muscles which rise in the foot, and each of which divides into a pair, as it crosses the trunk, and then subdivides into smaller branches, as it passes over the coronal cup to the base of the lobes. Mr. Gosse makes them to be five pairs, and says that usually each pair runs up the trunk from the foot in a line with one of the arms; and then, before reaching it, divides into diverging branches which, at remote points, are united to a muscular collar close to the base of the arms. He notices, however, that he has seen cases where the muscles run down direct from the depressions between the lobes without uniting to form pairs.

My own opinion, after prolonged observation of many specimens, is that there are really six pairs of muscles, and that they are arranged in the following fashion. Each pair runs up the foot looking like a single muscle; and the reason why never more than four (pairs) are visible in the foot from any point of view, is that there is always a pair on each side of the animal (however viewed) which is there lost to sight. At the junction of the foot and trunk each pair begins to open a little; and by the time they have reached the bottom of the coronal cup the constituents of each pair diverge obviously from each other, and terminate usually at the base of some one of the depressions between the lobes: but in such a fashion that the constituents of the same pair never end in the same depression (Pl. IV. figs. 2, 3, 4, *bm*). There is, however, an exception to this in the case of the two pairs of dorsal muscles (Pl. IV. fig. 2, *bm*). Here it will be seen that while the outer muscles in each pair end in a depression between the lobes, the inner muscles curve over towards each other and meet so as to form a fine arch, some distance below the base of the dorsal lobe. There are, too, fine hexagonal markings visible on this side of the coronal cup, which are probably the boundaries of large cells: oval nucleated cells are also easily seen in the wall of the coronal cup, when the animal is viewed from either side (fig. 4).

The **nutritive and reproductive** systems are so similar to those of *Floscularia*, that they require no separate description. It is enough to call attention to Dr. Leydig's figure of the ovary treated with acetic acid (reproduced in fig. 7), and exhibiting the ova in various stages of growth, as well as its own delicate walls, and the oviduct (*ot*), which leads into the cloaca (*cl*).

The **Secreting System.**—Neither salivary, gastric, nor foot glands have been observed in *Stephanoceros*, but as the animal secretes a large and comparatively solid tube, it is clear that it must either have some organ for this purpose, or that the substance of which the tube is constructed oozes from the surface of the body.

The **vascular system** is much better seen in this genus than it is in the preceding. Fig. 4 shows *Stephanoceros* viewed a little obliquely from the side on the left of the dorsal surface. The left lateral canal (fig. 4, *lc*) can be seen winding to the left of the nervous ganglion (*gn*) and having two vibratile tags (*rt₁*, *rt₂*) attached to it close to where the left eye (*e*) is. The lateral canal then divides into two branches; the right branch curving upwards towards the dorsal surface to meet its fellow on the median dorsal line (see fig. 2), while the left branch passes along the side of the vestibule till it nearly reaches the level of the knobbed ends of the ciliary wreath (c.c). Here this left branch joins an offshoot of the right branch; the point of junction being marked by a third vibratile tag (vt₃). A fourth vibratile tag (vt₁) is attached to the right branch just where it gives off the connecting offshoot, and a fifth (vt₂) can be seen on the highest dorsal portion of the lateral canal. Fig. 2 shows the same right and left canals, as seen from the dorsal surface, with the same vibratile tags on either side. Each lateral canal winds down the side of the trunk and ends at last on the surface of the contractile vesicle. Leydig (loc. cit.) records his having distinctly seen this junction in young specimens, as well as a duct leading from the contractile vesicle into the cloaca (cl).

The **Nervous System.** What is probably the **nervous ganglion** is a peculiar organ (figs. 2, 4, 5, gn) consisting of large clear cells, lying at the back of the vestibule near the dorsal surface. Above it, and well under the dorsal surface, is a three-lobed, granular, semiopaque body (figs. 2, 5, x) with which the nervous ganglion is possibly connected. The nervous ganglion in many of the Rotifera, especially among the *Notommatadæ*, shows a marked cellular structure at the lower end which loses itself in a granular, semiopaque upper portion; but it must be admitted that if these peculiar bodies (gn, x) constitute the nervous ganglion of *Stephanoceros*, it is rather their position than their shape and structure that would lead us thus to interpret them. From the spot where it adheres to the wall of the vestibule, a sort of protrusile tongue or **taster** (fig. 1, tr) rises which can be pushed forward so as nearly to fill up the interval between the knobbed ciliated ends of the ciliary wreath. This tongue may be seen incessantly pressing backwards and forwards as the food passes into the vestibule, as if discriminating between the passing atoms, just as the two tasters do in *M. ringens.*[1] The **eyes** (fig. 2 e) lie on either side of the nervous ganglion; they may be seen by dark-field illumination, but as they are small, rather deep down under the surface, and often obscured by other parts, it is not easy to get both into view at once. Mr. Cubitt (loc. cit.) describes and figures them as clear globes resting on pigment spots, and with nerve threads attaching them to the nervous ganglion: this is a very probable structure, but I have failed to make it out. Two very short lateral **antennæ** (Figs. 2, 5, a) can be seen when *Stephanoceros* is viewed dorsally: they are mere setigerous pimples.

The **development of the young** is shown in Mr. Gosse's figures (Pl. IV. figs. 8 to 15), in which fig. 8 represents young *Stephanoceros* a few minutes after birth; figs. 9 and 10, a little later; and figs. 11 to 14 represent successive stages of growth of a specimen from three to eighteen hours old. Fig. 15 shows the perfectly developed young *Stephanoceros*, thirty-six hours old; it exhibits the bands of setæ, the principal viscera, the muscular collars, vestibule, crop, and jaws. Mr. Rosseter (loc. cit.) says that on one occasion he watched the development of a young *Stephanoceros* and noticed that the lobes of the corona "originate as buds and unroll like the fronds of ferns" (figs. 16, 17). These buds began to appear about eleven hours after the animal was hatched, and when they had risen to a small height gradually unfolded; they remained in a drooping state for two days, but on the third day took the arched form usual in the adult. Dr. Mantell observed a young specimen in which the lobes even after eighty hours from birth were mere rudimentary buds. Such discrepancies in the rate of development noticed by these three observers are common in all the Rotifera, and are doubtless partly due to the various degrees of development that the embryo attains in the ovum before its extrusion. In *Stephanoceros* (as in a few other Rotifera) the young (as Ehrenberg conjectured) is occasionally born alive. This has been seen by Mr. Rosseter and Dr. English,[2] and indeed is almost shown in Dr. Leydig's figure (Pl. IV. fig. 7), where the much advanced embryo (y) lying close to the oviduct (ot) already exhibits the eyes and frontal cilia.

---

[1] (Pl. V. fig. 2 (tr).   [2] See Mr. Rosseter's paper (loc. cit. p. 171).

No **male** has as yet been recorded, but Leydig's fig. 3, Taf. i. (*loc. cit.*), of a young *Stephanoceros* forced out of the egg by pressure, has a very masculine look about it.

Disease. – Mr. Gosse (" Popular Sci. Rev." *loc. cit.*) has noticed that *Stephanoceros* will occasionally throw off portions of one or more of its lobes, which slough away so as to be reduced to mere stumps. As Mr. Gosse remarks, there is little doubt that Perty's *S. glacialis* (*loc. cit.*) is only such an unhealthy specimen of *S. Eichhornii*. I have seen the same thing ; and I have noticed that, when a portion of the lobe sloughed off, the discarded piece would round itself into a globe, and float away urged by the fitful lashing of the setæ on it.[1]

Length, $\frac{1}{24}$ to $\frac{1}{12}$ inch. **Habitat.** On weeds in clear ponds in the neighbourhood of London (P.H.G.) ; of Birmingham (T.B.) ; in marsh pools on *Sphagnum*, Perthshire (J.H.). Very partially distributed ; not uncommon about London and Birmingham, but rare in Scotland, and very rare apparently in America. Widely spread on the Continent.

[1] Mr. Newlin Peirce (*loc. cit.*) has written a strange account of a *Stephanoceros* that accumulated a mass of *debris* in the upper portion of its tube ; and then, dividing itself transversely at the level of the *debris*, deserted its tube, carrying the accumulation with it, and attached itself to another stem of the plant to which it was originally fastened. Here it gradually became a perfect animal in a new tube ; and it then repeated the process. The whole account is incomprehensible.

CHAPTER VI.

MELICERTADÆ.

Chaque genre de Vers, et j'ose presque dire chaque espèce, offre un objet tout à fait neuf, qui demande à lui seul presqu'autant de travail que les classes entières des grands animaux.—GEOFFROY ST. HILAIRE.

What, dull! when you do not know what gives its loveliness of form to the lily, its depth of colour to the violet, its fragrance to the rose! when you do not know in what consists the venom of the adder, any more than you can imitate the glad movements of the dove! when, unlike the wisest of monarchs and of men, far from knowing the trees as he did, " from the cedar tree that is in Lebanon even unto the hyssop that springeth out of the wall," you do not know anything even of the two extremes of Solomon's great knowledge! What, dull! when earth, air, and water are all alike mysteries to you! and when, as you stretch out your hand, you do not touch anything the properties of which you have mastered! while, all the time, Nature is inviting you to talk earnestly with her, to understand her, to subdue her, and to be blessed by her! Go away, man; learn something, do something, understand something, and let me hear no more of your dulness.—SIR ARTHUR HELPS.

CHAPTER VI.

Family II. MELICERTADÆ.

Corona *not produced into setigerous lobes ;* **buccal orifice** *lateral ;* **ciliary wreath** *a marginal continuous curve, bent on itself at the dorsal*[1] *surface, so as to encircle the corona twice, with the buccal orifice between its upper and lower curves, and having also a dorsal gap between its points of flexure ;* **trophi** *malleo-ramate.*

The *Melicertadæ* are at once distinguished from the *Flosculariadæ* by the difference of the corona, and the unsymmetrical position of the buccal orifice. In all the genera the corona bears two parallel wreaths of cilia, the upper of which frequently presents the appearance of a revolving wheel. The family contains seven genera, which differ from each other mainly in their coronæ, tubes, and habits ; their internal structure being so much alike, that it has been proposed, more than once to reduce the seven genera to two.

There is no more interesting family. It contains animals that build their own tubes, pellet by pellet ; and that themselves form these pellets, either out of external materials, moulded in hollows of their own bodies, or out of their own fæces. All have social instincts : some rearing their tubes, to the fourth and fifth generation, on those of their ancestors, or forming dense clusters on the stems of water-plants ; and others (fixed forms only in a sort of Parliamentary sense) adhering to each other by their posterior extremities, and forming spherical clusters that roll unceasingly through the waters of still lakes and ponds. Most of them are hardy, and luckily all are prolific ; sometimes so amazingly that the water-weeds are literally covered with their tubes, and the fortunate finder can thus have in the small compass of a live box scores of animals of all ages, and in every stage of growth.

Genus MELICERTA.

GEN. CII. **Corona** *of four lobes ;* **dorsal gap** *wide ;* **dorsal antenna** *minute ;* **ventral antennæ** *obvious.*

The **tube** varies in all the four species, and its structure and formation will be described under each. In all there is an inner gelatinous tube,[2] and in *M. ringens* and *M. coniifera* there is also an outer tube, consisting of pellets of extraneous matter ; while in *M. Janus* the pellets are fæcal. In *M. tubicolaria* the outer tube is entirely absent.

The **corona** seen dorsally looks somewhat like a heart's-ease, with its four petals lying in a plane ; but a side view shows that the two lower lobes are bent upwards, so as to form an oblique angle with the upper lobes. A groove runs round the corona, on both sides, just under its edge ; and on the ventral surface it is confluent with the buccal funnel. There is a gap in the groove on the dorsal surface, so that it does not entirely surround the corona. The edge of the corona is fringed with large cilia, and the edges of the groove and buccal funnel with much smaller ones ; and they

---

[1] In one instance (that of *Conochilus volvox*) read ventral for dorsal.

[2] This inner tube can be seen in the young animal (Pl. V. fig. 1*d* and Pl. VI. fig. 1*g*) before the outer tube has been completed.

are all in constant motion. The action of the former, or principal, wreath (Pl. V. figs. 2c, 4, pw) draws the particles floating in the water into two spiral currents, which are tangential to the groove on either side of the disk. The action of the latter, or secondary wreath (Pl. V. figs. 2c, 4, sw) drags the particles, as in their spiral path they strike the groove, out of that path into the groove itself, and hurries them along its windings towards the buccal funnel. At the two points where the groove on each side joins the buccal funnel are two fleshy knobs (Pl. V. fig. 2c, tr), which can be seen in constant motion ; either regulating the force of the current, examining the moving atoms, or possibly performing both offices at the same time. The ciliated edges and sides of the buccal funnel conduct a portion of the stream down to the mastax ; while another, and apparently the larger portion, rushes over the ciliated chin.[1] If a little carmine be added to the water, it is a pretty sight to see the coloured spirals form on both sides of the corona, while two processions of crimson atoms wind in and out under the margin of the petals, one on the left hand, and one on the right ; each starting from the dorsal gap, and dashing at last down the buccal funnel, or over the chin. (Pl. V. figs. 2c, 4, ch.) But the whole of the ciliary apparatus has not yet been noticed. Beneath the chin there is a hemispherical hollow (Pl. V. figs. 2c, 4,cc) which is furred with minute cilia. It is in this cup that *M. ringens* and *M. conifera* mould their pellets, as will be described further on[2] ; it is present also in *M. tubicolaria* and *M. Janus*, but its function in these species is unknown.

The **Nutritive System.**—The buccal funnel (Pl. V. figs. 2c, 2d, 4, bf) slopes downwards, and somewhat dorsally, towards the mastax. It is ciliated throughout, and has a pair of chitinous lips (fig. 2d, lp) similar to those described at p. 6. These lips are in frequent motion, now opening and shutting, now moving up and down the funnel, evidently selecting and rejecting the food : if an objectionable morsel attempts to pass, " it is astonishing to see how the little quick jerk, which the lips give, tosses it up into the central stream of waste and drives it away."[3] Should this morsel be unusually large, the mastax itself aids the lips in their upward jerk. On each side of the buccal funnel and above the mastax is a clear organ (Pl. V. fig. 2d, sg) whose surface is spheroidal. The two have been described as salivary glands by some observers, and as mere stays to the mastax by others. They are obviously elastic, and move up and down with its every motion. The mastax (fig. 2d, mx) consists of three confluent lobes, presenting a trefoil outline in vertical section ; each side-lobe contains a malleus, and the bottom-lobe grasps the incus. The malleo-ramate trophi (Pl. V. fig. 1f) are almost precisely similar to those already described.

The food flows between the lips, and after having been torn by the sharp teeth of the mallei, and crushed by the ridged inner surfaces of the rami, passes through a short ciliated œsophagus (fig. 4, œ), and so enters the stomach (fig. 4, s). This is a long cylindrical sac, with very thick walls of large cells, lined with cilia. A partial constriction usually separates it from the intestine (i). The walls of this latter are thinner and more transparent, and their cilia longer. The distinction between the stomach and intestine is obliterated when the stomach is much distended with food, but is usually present, and is often rendered obvious by the difference of colour in the contents of the two. Those of the intestine revolve under the action of its cilia ; and when the pellet thus formed is ready for extrusion, the animal lifts its cloaca above the rim of its cup, pushes up the pellet, bends downwards over it, and then dexterously shoots it across its shoulder into the current flowing from the chin. By this means the fæces are carried away out of the currents of the trochal disc.

The **Secreting System.**— The so-called salivary glands I have already noticed. There

---

[1] Judge Bedwell (*Mon. Micr. J.* vol. xviii. 1877, p. 216) describes in *M. ringens* a hemispherical cushion, placed at an angle on that side of the buccal funnel which is opposite to the chin. He thinks that it is a highly sensitive organ, which, by altering its facial configuration, directs the streams that go down the buccal funnel and over the chin, and drives suitable particles in appropriate directions. I have not, however, been able to confirm these observations.

[2] Pp. 70, 71.   [3] Judge Bedwell (*loc. cit.*).

are two moderate-sized gastric glands (Pl. V. fig. 2d, gg) in the usual position at the top of the stomach; and Mr. Gosse describes in M. ringens "near the tip of the foot on its ventral side, a little granular body connected with the tip by a point, and enlarging at the upper end where it is connected with a small globular vesicle."[1] He suggests that this is a foot-gland similar to that in so many other Rotifera.

The **Vascular System.**—The contractile vesicle is very small, and is generally hidden by the viscera; it lies close to the rectum. When the animal is viewed sidewise, the lateral canals can be traced from a knot of twisted tubes in the shoulder to a similar knot in the corona. Two vibratile tags can be seen at each knot. The lateral canals lie close to the surface, and can be best found by slowly focussing upwards from any point near the shoulder, and just under the cuticle. It is possible to trace them down from the shoulder to the contractile vesicle; but it is not often that the viscera lying beneath permit this to be done.

The **Reproductive System.**—The ovary is similar in structure to those already described; it is somewhat oblong in shape, and extends between the stomach and ventral surface, over nearly the whole of the latter. The oviduct passes beneath the intestine, and in M. ringens " enters the cloaca near the point where the lower stomach [intestine] opens into the excretory canal."[2]

The **Nervous System and Organs of Sense.**—The nervous ganglion has as yet only been made out in M. ringens. M. Joliet (loc. cit. M. ringens) describes it as a group of big cells of a very characteristic form, and provided with a large nucleus. Many similar cells are placed beside the first, and stretch in different directions. It is not large, and is situated on the dorsal face of the pharynx. The two ventral antennæ (figs. 2c and 4, a) are very obvious. Between them lies the buccal funnel, as well as the tract which, in M. ringens and M. conifera, is engaged in forming and depositing the pellets; and which contains the chin (ch), the ciliated cup (cc), and a prominent knob lying just under it. The antennæ are tubes, with a short transverse plug in them, carrying a pencil of delicate setæ. This plug can be withdrawn into the tube at will, by a muscular thread; the top of the tube itself following the plug, just as in a snail's horn. They are so situated that, when Melicerta furls its corona, they stand on the top of the round closed head (fig. 1c, a). Immediately opposite to them, in M. ringens and M. conifera, there are two curved sharp hooks (h), which look like weapons of defence; and between them lies the third antenna, the dorsal one, which is nothing but a setigerous pimple. Two red eyes are visible in the young, but none as yet have been seen in the adult.

The **Muscular System.**—The longitudinal muscles, as in the *Flosculariadæ*, run up the foot to its junction with the trunk, where they are fastened. They then cross the trunk till they reach the neck, where they are again fastened; and as they reach the head they divide into branches, which cross the lobes of the corona, and, by their contraction, furl it. Transverse muscles, imbedded in the integuments, encircle the trunk; and, by the compression of the body-fluids, drive out and unfurl the corona, just as in *Floscularia*. In M. conifera I have observed a set of longitudinal branching muscles, which are inserted in the neck, just under the ciliated cup, and the prominent knob beneath it (Pl. V. fig. 2c, lm). It is by their varied action that *Melicerta* is able to curve and twist its neck, and clinch its pellet on to the top of the tube, by the opposing pressures of the knob and chin.

The **male** has not yet been certainly recognised. I have seen what is probably the male of M. tubicolaria,[3] and Judge Bedwell gives a graphic description of the probable male of M. ringens,[4] and of its coquetting with the female; Mr. Gosse has also seen what there is little doubt was the male of M. conifera.[5]

[1] Quart. J. Micr. Sci. vol. i. 1853, p. 71, pl. ii. fig. 22.
[2] Professor Williamson, Quart. J. Micr. Sci. vol. i. 1853, p. 1.
[3] See p. 73 and Pl. V. fig. 3c.   [4] Midland Naturalist, vol. i. 1878, p. 245; see also p. 71.
[5] See p. 72.

M. RINGENS, *Schrank.*

(Pl. V. fig. 1.)

Melicerta ringens		Ehrenberg, Die Infus. 1838, p. 405, Taf. xlvi. fig. 3.
		Gosse, Trans. Micr. Soc. vol. iii. 1852, p. 58, pl. xii. figs. 1 1.
		„ Quart. J. Micr. Sci. vol. i. 1853, p. 71, pl. ii. figs. 12 27.
		„ Popular Sci. Rev. vol. i. 1862, p. 471, pl. xxvi. figs. a, b.
		Pritchard, Infusoria. 1861. p. 672. pls. xxxii., xxxvi., xxxvii.
		Williamson, Quart. J. Micr. Sci. vol. i. 1853, p. 1, pl. i. figs. 14-34.
		Leydig, Ueb. d. Bau. d. Räderth. 1851, p. 17.
		Claparède, Ann. Sci. Nat. Zool. 5 Sér. t. viii. 1867, pl. iii. figs. 1, 2.
		Cubitt, Mon. Micr. J. vol. v. 1871, p. 205, pls. lxxxiii., lxxxiv.
		„ Mon. Micr. J. vol. viii. 1872. p. 8, pl. xxiii. fig. 2.
		Bedwell, Mon. Micr. J. vol. xviii. 1877, p. 211, pls. cxcvii. and cxcviii.
		„ J. Roy. Micr. Soc. vol. i. 1878, p. 176, pl. x.
		„ Mid. Nat. vol. i. 1878, p. 245.
		Hudson, J. Roy. Micr. Soc. vol. ii. 1879, p. 6.
		Joliet, Comptes Rendus, t. 93, 1881, pp. 748, 856.

SP. CII. **Lobes** *when expanded, wider than the tube;* **chin** *short, extremity blunt;* **pellet** *nearly spherical.*

During the hundred and eighty years which have elapsed since Leuwenhoek discovered *M. ringens,* it has been a source of delight to a long succession of observers. It has had more than a dozen names given to it, and has been the subject of upwards of three times as many treatises; and no wonder, for the surprising spectacle of its whirling disk captivates even those who have seen it scores of times before.

Then the building of its **tube** appeals powerfully to the imagination. Here is a tiny creature which, when barely an hour old, and not $\frac{1}{100}$ of an inch in length, sweeps from the water its food and the materials for its dwelling; and which, at the same moment, and with faultless accuracy, sorts the one from the other, and both from the mere rubbish, drives away the waste, sends a stream of food down its throat, supplies selected atoms to a brick-making machine in its own body, mixes them with cement, moulds them into bricks, and finally (to crown the marvel) lays the bricks one by one around its body in regular order, so as to form a compact and effective dwelling.

Leuwenhoek saw, and clearly described, the laying of the pellets and the raising of the tube; but failed to see how the former were produced. Indeed, even what he did discover was forgotten, so that Mr. Gosse's paper "On the Architectural Instincts of *Melicerta ringens*" (in which the process of forming the pellets and tube was completely described) roused the greatest desire in all microscopists to see this marvel for themselves.

It is unnecessary for me to quote once more passages that are to be found in every text book, but I will continue the description (interrupted at p. 68) of the various streams that are set in motion by the ciliated borders of the corona, buccal funnel, and chin, and briefly tell how the pellets and tube are formed.

The main stream of waste, that rushes over the chin, has two feeble currents running under the somewhat incurved edges of the buccal funnel: at the side of its banks, if I may so say. Along these two currents float very minute atoms, at a comparatively gentle rate, while the larger particles dash along in the main stream. As the former glide along the banks of the buccal funnel they come to a notch on either side of the chin, over which they slip and are then drawn by the action of connecting lines of fine cilia into the ciliated cup, that lies close beneath the chin. This cup is nearly hemispherical in shape, and is furred with fine cilia. Soon after it has been emptied of one pellet, another begins to form in it, and a minute sphere of particles, generally of a yellowish-brown colour, is seen whirling in the centre of the cup. As this rapidly grows in size from the addition of fresh matter, it is easy to see in what direction it rotates, by means of the darker specks on its surface. If these are watched, it will be found that every now and then the rotation is reversed, and that this happens many times before the pellet is completed. It is needless to credit *Melicerta* with the voluntary alteration of the motion, for it is obvious that a pellet, kept in the centre of a ciliated cup by the action of its cilia lashing up and down, must be in an unstable position ; a very little alteration of its own figure, or of its centre of gravity, or of the relative power of the cilia in different parts of the cup, would be sure to drive it out of its central position to one side or the other. This done, the cilia on that side (say the lower one) would be checked, and those on the upper would have the predominance, and so force the pellet to rotate towards the upper side ; which when it had slowly reached, the upper cilia would in their turn be checked, and the lower cilia would now have the predominance, and would again draw the pellet towards themselves, reversing the rotation —and so on. The particles in the cup are made to adhere by being mixed with the same glutinous secretion as that which forms the inner tube. This exudes either from the cup itself or (as I believe) from the surface of the large knob just beneath it (Pl. V. fig. 2c). In a minute or two, from the commencement of the process, the pellet is completed, and then the animal bends its neck swiftly over the edge of the cup, and clinches the pellet on to the top of the inner tube, by the opposing action of the chin and the knob beneath. It is obvious that it selects the place in which to deposit the pellet, and it is probably guided to the exact spot by its dorsal antenna, which is generally close to the spot the instant before the pellet is laid.[1] It is curious that Ehrenberg should have completely missed the way in which the pellets and tube are formed. He says that the former " are not foreign bodies (as in the tube of *Phryganea*) nor excrement ; but a peculiar substance mixed with the latter, gummy, and hardening in water " : and he further says that he distinctly saw the pellet discarded from the posterior intestinal opening, and fastened by it to the tube. Mr. Gosse, who calls attention to this discrepancy, suggests that there may possibly be " two species closely allied but differing in this part of their organisation and economy " ; and the discovery of *M. Janus*, which has precisely the habit wrongly ascribed to *M. ringens*, shows how shrewd was Mr. Gosse's suggestion. The only difficulty about the matter is that Ehrenberg's drawings are certainly taken from *M. ringens* ; while his description of the formation of the pellets and tube seems to be taken from *M. Janus*. Possibly he may have found first the one, and then the other, and not distinguished between them ; though that seems hardly likely.

Melicerta ringens in England does not usually occur in clusters of adhering individuals, though occasionally one is seen with a young one or two attached to its tube. But in the United States (as I have already noticed) it frequently occurs in large clusters, and some of the tubes of these clusters greatly exceed in size the largest known English specimen. For instance, I possess a cluster in which the central tube is ¼ of an inch long, and of which therefore the tenant must have been upwards of ¼ of an inch in length, thus exceeding even the great length of *Floscularia Hoodii*. I found that this great tube contained upwards of six thousand pellets arranged in about two hundred and forty rows, one above another.

The **Male.**—Judge Bedwell in the " Midland Naturalist " (*loc. cit.*) describes a small free-swimming Rotiferon which he saw emerge from a tube of *Melicerta ringens*. It was not more than ¼ of the length of the tube, had a forked foot, and trophi somewhat like an inverted W, which were capable of protusion through the corona. Like the probable male of *M. conifera*,[2] " it began to woo and caress the lobes of the female in the most active and elegant manner, almost as if it were nibbling the main wreath of cilia. Now to anyone accustomed to watch *Melicerta*, it must always be a matter of astonishment to see such a timid, nervous rotifer allow another to touch the cilia with impunity : but in this instance the female never flinched in any way, but accepted the attentions of

[1] Judge Bedwell, *Mon. Micr. J.* (*loc. cit.*). The whole paper is most suggestive.
[2] See Mr. Gosse's description, p. 72.

the little visitor with perfect composure, and continued to feed as if quite undisturbed by its presence." The same observer broke up about fifty tubes of *M. ringens* in December, and procured ten specimens of the same small Rotiferon from them : in one case there were four males in a single tube.

In the above account, the presence of a forked foot, and of a mastax and trophi, and the fact that the latter were seen to be protruded from the corona, would naturally lead one to say that the Rotiferon recorded was rather some one of the *Notommatadæ* than a male *Melicerta*. On the other hand, its unresented action towards the female was precisely that noticed by Mr. Gosse in the case of *M. conifera* ; and the latter observer has also seen trophi in a very similar creature with similar habits, which he believes to be the male of *Limnias ceratophylli.*[1]

Length. Varies greatly. Average length of an adult tube about ¼ inch. Specimens twice the size are common in Scotch lakes. Those in clusters, in Philadelphia, U.S., extend even to ½ inch. **Habitat.** Very common on water plants, in standing or slowly running water.

M. CONIFERA, *Hudson*, sp. nov.

(Pl. V. fig. 2.)

SP. CH. **Lobes,** *when expanded, of the same width as the tube ;* chin *long and pointed ;* pellet *a pointed cylinder.*

This *Melicerta* is somewhat larger, and very much rarer, than *M. ringens*. It was discovered by Mr. J. Hood in 1876 in a pool on Tent's Muir. He found it again in profusion in the summer, autumn, and even in some of the winter months of 1879, the weed being quite matted with it. The points of difference between it and *M. ringens* are persistent, though slight ; but the difference in their tubes is striking. This is due to the shape and quality of the pellets. They are much longer in proportion to their diameter than those of *M. ringens*, so as to resemble a conical rifle bullet : and they are more transparent, and of a clear golden yellow. In consequence of their length the tube is a stout one, and its thickness is shown by a stripe on each side of a different colour from the centre of the tube, and darker or lighter according to the illumination used.

The fully expanded lobes are almost exactly as wide as the top of the tube, but in *M. ringens* they exceed it in the proportion of ten to nine. The chin, too, differs from that of the common species : it is longer and more pointed.

The **Male.**—[In water from Epping Forest sent to me by Mr. Henry Davis. I found *Melicerta conifera*, projected and rotating. Emerging from the mouth of the tube, about three fourths extruded, was a male (Pl. D. fig. 6) about as long as the diameter of the tube, playing, as it were, with the disk of the female. Two irregular shaped opaque masses were seen in it far apart from each other. I looked away for a minute to delineate what I had seen, and he was gone : but I presently found him slowly swimming around, which he continued to do, turning on his long axis as he went. There was now only one opaque mass, the hinder ; and this was in contact (whether in connection I do not know) with a large ovate clear bladder, perhaps an air vesicle. The head is oblique, the face ciliated, the occiput, angled and projecting. The foot is a little knob of flesh. I could see no internal organs, nothing but the clear, colourless tissue, full of corrugations throughout. P.H.G.

Length. About 1/12 inch ; tube, 1/10 inch. **Habitat.** Marsh pools, Fife and Perth J.H. ; abundant in a pool at Snaresbrook (P.H.G.): rare.

M. TUBICOLARIA, *Ehrenberg.*

(Pl. V. fig. 3.)

Tubicolaria Najas .　　　　　Ehrenberg. *Die Inf.* s. 1838, p. 389, Taf. xlv. fig. 1.
　　　　"　　　"　　　　Leydig, *Ueb. d. Bau u. Einwick*. 1854, p. 14, Taf. 1. f. 7.

[1] See p. 70.

Tubicolaria Naias .

Melicerta tyro . · .

Tubicolaria Naias .

Pritchard, *Infusoria*, 1861, p. 668, pl. xxxii. figs. 379 382.

Hudson, *Mon. Micr. J.* vol. xiv. 1875, p. 225, pl. cxix.

Fullagar, *J. Quekett Micr. Club*, vol. iv. pp. 182, 202, pls. xvi.–xviii.

Lobes, *when expanded, more than three times the width of the body;* **antennæ** *very long;* **tube** *a gelatinous sheath without pellets.*

Three striking peculiarities at once catch the eye in this beautiful *Melicerta*, viz. (1) the great size of the trochal disk; (2) the wonderful length, backward setting, swollen bases, and great flexibility of the ventral antennæ; and (3) its semi-transparent tube, which, though resembling that of the Floscules, has yet a character of its own.

The **tube,** unlike those of all the other species, bears no pellets at any time. It is of a loose fluffy texture, and extends from the surface almost up to the Rotiferon itself. There is a central hollow, little wider than the creature's body, up and down which the animal moves. Horizontal layers of diatoms, and other foreign bodies, cross the tube at irregular intervals, and mark the height, at which the tube then was, when they were entangled.

The **corona** is very large. It is more than three times the width of the body; thus considerably exceeding the proportions of *M. ringens* and *M. conifera.* Its four lobes are really all curved, just like those of *M. ringens,* and are often seen fully expanded and round; but the animal has a habit of bending the corners of the two upper lobes, so as to give the whole disk a butterfly shape.

There is a **ciliated cup** under the chin, just as in the two former species, but no pellets are formed in it. I examined it carefully several times, but I could find no lines of cilia between the chin and cup, such as Mr. Gosse observed in *M. ringens;* neither could I see the minute notches in the chin, through which, in that species, minute atoms slip from the buccal funnel, to be conducted by lines of fine cilia to the pellet cup.

The **muscular, nutritive, secreting,** and **reproductive systems,** are so similar to those of *M. ringens* as to call for no further remark.

Vascular System.—Leydig notices the absence of the **contractile vesicle,** and says that he traced the two **lateral canals** from the corona, where they originate in two coils, across the body, from the dorsal to the ventral surface, and so down to the cloaca. He says that, after repeated efforts, he at last saw the two **vibratile tags** in the corona. I was more fortunate, for I made out five vibratile tags on each side of the body, though I could not see exactly where the lateral canals ended. I could find no contractile vesicle, but on two occasions I noticed that the empty intestine became distended and very transparent, and then shot out its fluid contents through the cloaca. While this was going on, the passage from the stomach to the intestine was closed. It would thus appear as if the intestine itself was filled by the lateral canals, and discharged the office of the contractile vesicle. It will be seen subsequently that something similar takes place in other Rotifera.

Organs of Sense.—I have failed to discover any **eyes** in the adult, but they are very conspicuous in the young animal (Pl. V. fig. 3*b*). The **antennæ** are of great length, twice as long in proportion as those of *M. ringens;* and, as they are transparent, it is easy to see how the muscle that runs up the centre to the setigerous knob at the top can withdraw the knob, thus infolding the tube, till the knob comes right down to the base of the antenna. When the animal is contracted into its tube, the antennæ are closely pressed to its club-shaped body; and, as it rises, they, too, slowly rise from their recumbent position, while the perivisceral fluid, under the pressure of the transverse muscles, drives the knobs up the antennæ, and so completely extends them.

The **Male.**—I believe that the male Rotiferon drawn in Pl. V. fig. 3*c*, is the male of *M. tubicolaria.* I had a small piece of *Anacharis* with about a score of females attached to it, and while observing them I saw this young male circling round one of them. It was about $\frac{1}{40}$ inch long; but, owing doubtless to its having just been hatched, the skin was so granular and corrugated that I could not make out its whole structure. The nervous ganglion, sperm-sac (ss), and penis (p), were plainly visible, and I could see the motion of the spermatozoa, though not the individual spermatozoa themselves: neither could I make out the muscles nor the water vascular system. I did not see this creature hatched; still, as there were no other Rotifera present but *M. ringens* and *M. tyro*, it was certainly the male of one or the other.

The Rotiferon, I have little doubt, is Ehrenberg's *Tubicolaria Naias*. He formed the genus to receive a Melicertan that was destitute of eyes at all periods of its life, and lived in a gelatinous tube. But Ehrenberg points out that he has not seen the young, and that therefore the absence of eyes cannot be depended upon as a characteristic. His other characters of the genus are, a four-lobed corona, two antennæ, and a gelatinous tube. Of these, the latter alone is peculiar to *Tubicolaria*, and as it is not sufficient to found a genus on, I have placed the animal among the *Melicertæ*.[1]

Length. Adults from about $\frac{3}{50}$ to $\frac{1}{10}$ inch; the maximum size given is that of Scotch specimens. **Habitat.** Ponds and ditches, Birmingham (C.T.H.); Chartham, Kent (Col. Horsley); Forfar (J.H.); Reading (Tatem): rare.

M. JANUS, *Hudson.*

(Pl. VII. fig. 1.)

Œcistes Janus   Hudson, *J. Roy. Micr. Soc.* 2 Ser. vol. i. 1881, p. 1, pl. i.

Upper lobes *deeply divided.* **lower** *nearly confluent;* **dorsal gap** *minute;* **antennæ** *short;* **chin** *two-pointed;* **pellets** *fæcal.*

This remarkable Rotiferon is one of those creatures whose form is as irritating to the classifier as it is delightful to the naturalist. For it possesses in almost equal proportions the characteristics of two genera, viz. of *Melicerta* and of *Œcistes*, and might, with nearly equal propriety, be placed in either genus. It was found first by Mr. J. Hood in Loch Laundie in 1880, and was most numerous, and in best condition, on weeds at a depth varying from six to ten feet.

When seen from the ventral surface, so that the lower lobes are partly hidden by its tube, no one would suppose it other than a *Melicerta*; but when it turns and exhibits its dorsal surface, it is seen that the lower portion of the **corona** resembles that of *Œcistes*, for, instead of there being a wide dorsal gap in the ciliary wreath, there is scarcely any at all (Pl. VII. fig. 1); and the subdivision of the lower portion of the corona is so slight that the outline of its two lobes is almost confluent: in fact, it might almost be said that this is a three-lobed Melicertan. As in *Œcistes*, thickenings (fig. 1d) cross the corona, which itself is so thin that it becomes nearly invisible under dark-field illumination, while the thicker portions stand out distinctly, especially when seen side-wise (fig. 1b). When the animal begins to open its corona, these portions are thrust forward in a squarish and very characteristic bundle, the thinner parts of the disk lying folded neatly between them. In this respect *M. Janus* closely resembles *Œcistes umbella.*

The cilia of the corona are unusually large, while the groove that lies between the principal and secondary wreaths is broad and deep. Should the larger cilia be checked by contact with the side of the cell in which the animal is placed, it is easy to count them, and their whip-like action becomes plainly visible. Individual cilia may now and then be seen even in the secondary wreath.

The chin (fig. 1b, ch) is also peculiar. It terminates, not in one point, but in two

---

[1] It must be admitted that Ehrenberg's figure is very unlike mine. The corona is barely the width of the body, and the antennæ are very short; but I think that both disk and antennæ are intended to be represented in a contracted state.

(fig. 1e); and below it, at right-angles to it, are two thin walls (fig. 1a) looking like the supports of a bracket, the chin being the bracket itself. These supports form, with the chin above, and the ventral surface between them, a cup or recess somewhat like the cup of *M. ringens*. I could not, however, detect any cilia in it. Below the cup is a large viscous knob, as in the other species, but differing from them in bearing on its upper surface two or more curved bristles (fig. 1b, h) pointing to the cup.

The **tube** is not nearly so compact as that of *M. ringens*, or even of *Limnias ceratophylli*. It is composed of large ovoid fæcal pellets, which are laid upon one another somewhat obliquely in rings, as shown in fig. 2d.

The pellet is formed in the intestine (figs. 1b, 1c, i), and when it is ready the animal bends down over the cloaca (cl), the rectum (r) is everted and pushed forward, and the pellet is shot over the shoulder so that it rubs against the viscous knob and is thus held, partly by the sticky surface of the knob, partly by the curved bristles already mentioned. When it has thus caught the pellet, the creature, with a swift twist backwards, pushes it on to the top of its tube. The **intestine** (i) is large and usually contains a fæcal pellet in the course of formation. These are not all appropriated to the construction of the tube, but are often suffered to float away. The **gastric glands** (gg) are conspicuous, and form an arch over the top of the stomach : they contain large nucleated cells. The two ventral **antennæ** (a) are short and are placed rather as they are in an *Œcistes* than in a *Melicerta*. Mr. Hood tells me that he has seen the **male**, and that it resembles that which I have figured as the male of *M. tubicolaria*. He further notices that the young **female** acquires its perfect form in four days after it has been hatched. Although *M. Janus* has so many points in common with the genus *Œcistes* that I originally placed it there, yet as it really has four lobes in its corona (however inconspicuous two of them may be) I have felt constrained to add it to the *Melicertæ*.

Length, ¹ₗ of an inch. **Habitat**. Lochs only, Forfar, Perth, Ayr (J.H.) : abundant.

Genus LIMNIAS, *Schrank*.

GEN. CII. **Corona** *distinctly of two lobes ;* **dorsal gap** *wide ;* **dorsal antenna** *minute ;* **ventral antennæ** *obvious ;* **tube** *without pellets.*

The genus *Limnias* differs from *Melicerta* in the shape of its lobes, and the structure of its tube. The **corona** is much broader than it is high, and consists of two nearly circular lobes connected on the ventral side by a hollow opposite to the buccal funnel, and separated on the dorsal side by a gap. The double **ciliary wreath**, buccal funnel, and chin are similar to those of *Melicerta*. There is no ciliated cup, though there is a hollow beneath the chin somewhat like one.

The **tube** is unlike that of *Melicerta* or *Œcistes* ; it is really tubular in form, widening a little towards the top ; it is often opaque, and is not composed of pellets ; those of the different species are unlike each other.

The **internal structure** of the common species, *L. ceratophylli*, is almost precisely that of *Melicerta* : it is said by Ehrenberg, and often repeated, that it has no vascular **system**, but this is a mistake. I have seen, but have not had an opportunity of studying, the rare species *L. annulatus* ; neither have I been able to find any account of its internal structure.

L. CERATOPHYLLI, *Schrank*.

(Pl. VI. fig. 1.)

Limnias ceratophylli	Ehrenberg, *Die Infus*. 1838, p. 402, Taf. xlvi. fig. 1.
„ „	Gosse, *Evenings at the Microscope*, 1859, p. 302, with fig.
„ „	Pritchard, *Infusoria*, 1861, pl. xxxii. figs. 388 392, pl. xxxvi. fig. 2.
Melicerta ceratophylli	Gosse, *Popular Sci. Rev.* vol. i. 1862, p. 181, pl. xxvi. fig. c.
Limnias ceratophylli	Moxon, *Trans. Linn. Soc.* vol. xxiv. 1864, p. 158, pl. xlvii. fi 8.

Limnias ceratophylli. Tatem, *J. Quekett Micr. Club*, vol. i. 1868, p. 121, pl. vi. figs. 1, 2.

L. cras... ni... Leidy, *Proc. Acad. Nat. Sci. Pa.* 1874, p. 110.

Limnias ceratophylli Bedwell, *Mon. Micr. J.* vol. xviii. 1877, p. 221, pl. cxeviii. figs. 7, 8.

SP. CH. *No horny* **processes** *on the dorsal surface below the corona ;* **ventral antennæ** *very short ;* **tube** *nearly cylindrical, smooth, often rendered opaque by extraneous materials, except at the posterior end.*

I have already related, in Chapter I., Leuwenhoek's discovery of this the earliest known tube-maker. It has not been much studied, as its tube is often quite opaque, and its own attractions have been eclipsed by those of *Melicerta*. The tube is of a yellow-brown tint and is generally coated over on the outside with waste matter that falls down on it from the coronal currents above, and with the particles that trickle over the chin, and adhere to the sticky surface beneath it. These latter are rubbed off from time to time on to the tube by the animal, as it bends its head over it. Doubtless this renders the tube smooth and compact. Judge Bedwell (*loc. cit.*) thinks that there is a chitinous shield below the dorsal gap, whose hard edge is shown at *xx.* fig. 1*d*. He points out that its position corresponds to those of the horny processes of *L. annulatus*, and the sharp hooks of *M. ringens*; and he suggests that the tube is smoothed with it "much as a bricklayer smooths over his stucco with his flat trowel." The tube is generally not coated towards its posterior extremity, and is very imperfectly covered in the young (fig. 1*g*). Occasionally adults are met with that have tolerably transparent tubes,[1] and even large adults have sometimes tubes of an opaque white (fig. 1*b*).

Ehrenberg recognises no vascular system,[2] but Dr. Moxon (*loc. cit.*) has observed part of it, and given a figure of the neck and expanded corona, with two **vibratile tags** on the same side. I have had no difficulty in seeing the **lateral canals,** and their accompanying tags, in the upper portion of their course, from a vascular plexus near the shoulder, up to a similar one in the corona (fig. 1*f*, *lc*, *et*). The **contractile vesicle** (if there is one) has not yet been noticed.

Besides the two short ventral **antennæ** (Pl. VI. figs. 1*d*, 1*f*, *a*) Dr. Moxon (*loc. cit.*) has observed a minute dorsal one similarly situated to that in *M. ringens*.

Prof. Leidy (*loc. cit.*) says that in many localities of the Schuylkill, almost every stone exhibits multitudes of bunches of a *Limnias*, pendent from its sides, and under surface; as many as fifty tubes may be counted in a bunch. Prof. Leidy proposes to call this rotifer *L. socialis*, on account of its habit of growing in clusters; but as the animal itself is said to be like *L. ceratophylli* in other respects, and as *L. ceratophylli* in England has this habit of clustering to a considerable degree[3] it is, unnecessary to make a new species of the American Rotiferon.

Male. As a *Limnias* was slowly protruding from its tube, there swiftly pushed past it, out of the mouth of the tube, a young one, which I supposed from its general appearance, a male.[4] It was a simple cylinder of colourless flesh, slightly tapering behind to a blunt point, with no foot or tail apparent, of about one-third of the total length of the parent, filled with minute globules of oil or air. There was a simple crown of cilia around the truncate front, a well-sized and well-made mastax, an enormous blunt-pointed brain-sac reaching about two-thirds down its total length, and carrying, on its dorsal side, near its point, a small but clear round eye-spot of crimson hue. Its manners were those common to males, swimming swiftly around the parent, often coming close to her for a moment, and then darting finally off on a wide wild voyage. That this was truly a male individual of the species is highly probable, notwithstanding the presence of a mastax, of which there was no doubt, and of a long viscus below which appeared to be a stomach. P.H.G.

Length. Maximum about 1/8 inch. **Habitat.** On water plants: very common.

[1] Mr. Gosse (*loc. cit.*), p. 303.   [3] "Klumen und Gefässe sind nicht erkannt."
[2] p. 38. Pl. VI. fig. 1c.   [4] Pl. P. fig. 7.

Mr. Tatem (loc. cit.) described and figured a tube-making Rotiferon, somewhat resembling a *Limnias*, with long curved antennæ. If the bi-lobed condition of the trochal disk (as given in his figure) were permanent, and if there were also an obvious dorsal gap in the ciliary wreath, this would be a species of the genus. Unfortunately Mr. Tatem's description does not make this point clear, and in other respects the animal resembles *Œcistes longicornis*.

L. ANNULATUS, *Bailey.*

(Pl. VI. fig. 2.)

Limnias annulatus . . . . Cubitt, *Mon. Micr. J.* vol. vi. 1871, p. 165, pl. xcviii.

SP. CH. Five horny **processes** *on the dorsal surface below the corona ;* **ventral antennæ** *moderately long ;* **tube** *cylindrical, smooth, transparent, and crossed by transverse ridges, at regular intervals.*

I have met with this *Limnias* but once, and then had no time to study it. It appears to be rare in England, though Mr. Cubitt (from whose paper, cited above, the following particulars are taken) found it in abundance in a tributary of the Northbourne Brook, Kent.

Its **tube** is remarkable. In young adult specimens it is perfectly transparent, with a brilliant orange tint which at the sides becomes a deep carmine, owing to the greater thickness of tube which we there look through. Fine transverse ridges, one-ten-thousandth of an inch apart, surround the tube from top to bottom. These are possibly caused by the pressure against the case of two **horny processes** (figs *2b, 2d, hp*) on the dorsal surface. These, under dark-field illumination, show a bright red spot at the top. Three less prominent processes lie in a line below the first two ; the distance between the rings is precisely that between these two rows, so that possibly they may help to form and gauge them.

The **foot** is long, wrinkled, with distinct muscles. When the animal contracts itself into the case it folds the foot as shown in fig. 2a ; a plan also adopted at times by *L. ceratophylli.*

Length, 1/45 inch ; Scotch specimens, 1/15 inch. **Habitat.** Witlingham, Norwich (Brightwell) ; Kent (Cubitt) ; Forfar (J.H.) : rare.

Genus CEPHALOSIPHON, *Ehrenberg.*

GEN. CH. **Corona** *nearly circular ;* **dorsal gap** *distinct ;* **dorsal antenna** *obvious ;* **ventral antennæ** *absent [1] ;* *two* **dorsal hooks** *enclosing the dorsal antenna.*

C. LIMNIAS, *Ehrenberg.*

(Pl. VI. fig. 3.)

Cephalosiphon Limnias .	Ehrenberg, *Verhandl. der Berlin Akad.* 1853, p. 193.
,, ,, .	Pritchard, *Infusoria,* 1861, p. 670.
Limnias ceratophylli .	Slack, *Marvels of Pond Life,* 1861, p. 149, with fig.
Cephalosiphon Limnias .	,, *Intellectual Observer,* vol. i. 1862, p. 53, figs. 1, 2.
,, ,, .	Gosse, *Intellectual Observer,* vol. i. 1862. p. 19, with plate.
Melicerta Cephalosiphon .	,, *Popular Sci. Rev.* vol. i. 1862, p. 490.
	Tatem, *J. Quekett Micr. Club,* vol. i. 1868, p. 124,
	pl. vi. figs. 6, 7.
,, ,, . . .	Cubitt, *Mon. Micr. J.* vol. v. 1871, p. 170, pl. lxxxi. figs. 1, 2.
,, ,, . . .	Hudson, *Mon. Micr. J.* vol. xiv. 1875, p. 165, pl. cxvii. fig. 1.

SP. CH. **Dorsal antenna** *very long ;* **tube** *tapering to the foot, compact, strengthened with extraneous material.*

[1] I thought once that I caught sight of a minute setigerous ventrally placed pimple in *C. Limnias* ; there may possibly be a pair of such ventral antennæ.

Ehrenberg formed the genus *Cephalosiphon* of his family of the *Floscularicea* to receive a single species, *C. Limnias*, one specimen of which he found on *Ceratophyllum* at Berlin. His characters of the genus are as follows : " *Cephalosiphon*, E. Rotatory organ bilobed, eyes two, sheath or lorica single, two little frontal horns enclosing the siphon ; " and those of the species are : " *C. Limnias*, E. Sheath membranaceous, ringed."

The characters of the genus and species were given in Pritchard's " Infusoria " (edition 1861), but no one after Ehrenberg seems to have actually seen the animal itself, till Mr. Slack found it in a pond in the neighbourhood of London in 1860 on *Anacharis alsinastrum*. Mr. Slack supposed it to be the young of *Limnias ceratophylli*, and gave a brief description of it under that name in 1861 in his " Marvels of Pond Life " (loc. cit.). He noticed the creature's bi-lobed corona, as well as the great length, flexibility, and peculiar action of the dorsal antenna " thrust on this side, and on that, as if to collect information for its proprietor." Mr. Gosse in the same year, in a paper entitled " A Rotifer new to Britain (*Cephalosiphon Limnias*)," gave a full description with a plate of three figures of the new Rotiferon, taken from some specimens sent to him by Mr. Slack. These specimens seem to have been injured by the journey, as they did not expand freely, and so led Mr. Gosse to draw the corona with a butterfly-shape, which healthy specimens do not possess. Mr. Gosse, however, fully worked out the Rotiferon's structure, with the exception of the secreting and vascular systems ; and he described and figured the " frontal horns " or **hooks**, which are situated like the hooks of *Melicerta ringens*, one on each side of the dorsal antenna. This Rotiferon is very partially distributed. It was upwards of twenty years after I first began to search for Rotifera in the neighbourhood of Clifton, that I first lighted on it ; and Miss Saunders has had a similar experience at Cheltenham. In 1875 I found a group of them on a leaf of a *Potamogeton* in a pond at Nailsea, near Bristol, and I made a careful drawing of the group (Pl. VI. fig. 3). The **tube** is horn-shaped tapering to the foot ; generally neater and more compact than that of *Œcistes crystallinus*, but coated with much the same sort of yellow-brown material.

The **trunk** is small compared with the **foot**, which is long and slender. The animal arches its dorsal side (fig. 3), in a manner common among free-swimming Rotifera, but unique among the fixed ones, which, in all other instances, arch the ventral side, so as to bring the entrance to the buccal funnel uppermost. *C. Limnias* has that entrance almost hidden by the bending over of the corona. The arrangement of the double **ciliary** wreath is precisely that of the other *Melicertadæ*. The usual pair of clear vesicles (**salivary glands** ?) rest on each side of the top surface of the mastax, which is high in the neck towards the dorsal side ; and the ciliated buccal funnel slopes across to pass between them. There are a narrow œsophagus, two globular **gastric glands**, cylindrical **stomach**, short **intestine** with upturned **rectum**, ending in a **cloaca** rather low on the dorsal surface.

Of the **vascular system** nothing has been seen ; but Mr. Gosse (loc. cit. " Intell. Obser.") describes the **nervous ganglion** as " a grey cloudy mass of irregularly-lobed form, immediately below the antenna, and behind the discal mammilla." I thought once or twice I caught sight of a ventral setigerous pimple just below the entrance to the buccal funnel, but I am not sure about it : there may be a pair of them there. The dorsal **antenna** is the striking feature in *C. Limnias*. When the animal has closed its corona and retired into its case, this slender transparent rod, with a brush of setæ at the top may be seen gently moving about to see if the coast is clear. When satisfied that it may come up safely, *Cephalosiphon* hitches its long antenna over the side of the tube, and hoists itself up by it into a great curve ; it then straightens its body and unfurls its corona.

The long antenna is not always straight ; it is occasionally bent into long curves like the process of *F. cornuta*, but its changes of form are slow. Its base is broadened out like that of a rose-thorn, as if to give it a good purchase. Two red **eyes** are conspicuous in the adult, a little below the dorsal surface, one on each side of the antenna, and close to the nervous ganglion.

The gradual changes of form in the **young** are shown in figs. 3, *x*, *y*, *z*, which are all taken from the same individual in different stages of growth. Fig. 3, *x*, shows it when only $_2\frac{1}{0}$ inch long, and with a sort of hump where the dorsal antenna is to be. Twenty-four hours after, the hump had become a short antenna (fig. 3, *y*); in four days the young animal had grown to $_3\frac{1}{0}$ inch, in six days to $_2\frac{1}{0}$ inch, and in twelve to $_3\frac{1}{0}$ inch, by which time, as shown in fig. 3, *z*, the characteristic antenna was well developed: at this stage of its growth I unfortunately lost it.

I have little doubt that M. du Trochet was the first discoverer of *C. Limnias*. In the "Annales du Muséum d'Histoire Naturelle," t. xix. 1812, p. 385, pl. 18, figs. 19 to 21, he describes and figures a tube-maker, *Rotifer cruciger*, with a fawn-coloured tube and a long dorsal antenna, and he noticed that the animal explores with it in all directions. It is true that he figures two eyes. "saillants et globulaires," near the summit of the dorsal antenna, one on each side of it; but these (if the animal were *C. Limnias*) must have been something extraneous, accidentally attached to the antenna; possibly two air-bubbles.

Length. About $_1\frac{1}{0}$ inch. **Habitat.** Neighbourhood of London (Mr. Slack); Sandhurst, Berks (Dr. Collins); Woolston, near Southampton (P. H. G.); Cheltenham (Miss Saunders); Lochs, Forfar (J. H.); Nailsea, near Bristol (C. T. H.): very partially distributed.

C. CANDIDUS, *Hudson*, sp. nov.

SP. CII. Dorsal antenna *very long;* **tube** *irregular, semitransparent, gelatinous.*

Mr. J. Hood found this very rare Rotiferon in Loch Lundie in October 1880, and again between September and December in the same loch next year. He found a few specimens in his aquarium, but evidently from eggs on weeds brought from the same spot.

Mr. Hood points out that it differs from *C. Limnias* not only in its **tube**, but also in its stout wrinkled **foot**, which is kept habitually in this condition. I have always noticed that the foot of *C. Limnias* (when in its tube) is delicate, tapering, and smooth; but Mr. Gosse found that, on his dislodging one from its tube, the foot became of nearly equal thickness throughout (as in *C. candidus*), and of about one-third the diameter of the body, from which it was abruptly separated. The whole length, too, was then studded with wrinkles, which at intervals took the form of great thickened rings. The foot terminated in a circular sucking disk. The two species are very much alike; but their tubes differ so much that I have thought it best to keep them apart.

Length, $_1\frac{1}{0}$ inch. **Habitat.** Loch Lundie (J.H.): very rare.

Genus ŒCISTES, *Ehrenberg.*

GEN. CII. Corona *a wide oval, indistinctly two-lobed;* **dorsal gap** *minute;* **dorsal antenna** *absent* [1]; **ventral antennæ** *obvious.*

In this genus the tube is extremely irregular and variable in shape; it is usually of a loose fluffy texture, and encumbered with extraneous matters of all kinds. In one species, however, it is compact, like that of *Limnias ceratophylli;* in another it is formed of fæcal pellets; and in a third it resembles that of *Melicerta tubicolaria.*

The corona is no longer distinctly lobed, and becomes nearly circular; it is hollowed a little on the ventral side opposite to the buccal funnel. The ciliary **wreath** is double, and is on precisely the same plan as that of the preceding genera; only the dorsal gap in it, though distinct, is so minute as easily to escape notice, unless the animal be in a favourable position. Ehrenberg failed to see either it or the secondary wreath; and,

---

[1] Probably a minute setigerous pimple, which has escaped observation.

in consequence, separated it from the *Melicertidæ*, and placed it in a family named after itself: an honour it did not deserve. The antennæ vary greatly in length in the different species: some are mere setigerous pimples, others the longest known among the Rotifera.

The internal structure, so far as it has been ascertained, is that of *Melicerta*. The parts not made out are the vascular system and the nervous system. The male also is unknown.

Œ. CRYSTALLINUS, *Ehrenberg*.

(Pl. VII. fig. 3.)

Œcistes crystallinus Ehrenberg, *Die Infus.* 1838, p. 392, Taf. xliii. fig. 7.
" " Pritchard, *Infusoria*, 1861, p. 663, pl. xxv. figs. 361 364.
Melicerta crystallina Gosse, *Popular S. . Rev.* vol. i. 1852, p. 490.

SP. CH. Ventral antennæ *extremely short, and set wide apart;* tube *most variable and irregular in shape, often beset with extraneous matter.*

I have frequently met with this Rotifer in dense colonies, whose dirty-brown tubes gave quite a rusty look to the water plants which they infested. Though small, it is a very pretty object; and, with a little care, the secondary wreath of cilia, the two antennæ and the dorsal gap in the ciliary wreath can be distinctly made out. For this purpose the animal must be so placed that its long axis is nearly in the line of sight. From almost all other points of view the dorsal gap is invisible, and only one antenna can be seen at once, owing to the unusual width between the two. When the case is free from rubbish, it is not difficult to see that the animal's internal structure is very like that of *Melicerta*. The vascular system and the nervous system have not yet been observed. Two red eyes are visible in the half-grown young.

Length. About ¹⁄₁₄ inch. Habitat. Ponds and ditches: very common.

Œ. INTERMEDIUS, *Davis*.

(Pl. VII. fig. 5.)

Œcistes intermedius Davis, *Trans. Roy. Micr. Soc.* vol. xv. 1867, p. 11, with fig.

SP. CH. Antennæ *short;* dorsal gap *unusually wide;* tube *opaque, tapering slightly from top to bottom.*

Mr. Davis found this species at Leytonstone, in company with the former. It differs from *Œ. crystallinus* in the width of the dorsal gap in the ciliary wreath, which almost approaches that of a *Limnias*, and in its neat tube, which exactly reproduces that of *L. ceratophylli*. In fact, had it not been for its distinctly oval corona, I should have said that it was a variety of the latter species. Mr. Hood, however, tells me that he has found it (sometimes in abundance) in Scotland and always with the oval disk: Mr. Gosse, too, has seen many Scotch specimens, and has no doubt that it is a true species.

[Two specimens, so young that no visible tube was begun, yet attached to a stem by the foot, showed, in the wall of the occiput just below the ciliary rota, two well defined and conspicuous dark eyes, rather far apart.—P. H. G.]

Length. About ¹⁄₁₄ inch. Habitat. Leytonstone, Essex (H. Davis); marsh pools, Fife and Perth (J.H.): not common.

Œ. SERPENTINUS, *Gosse*, sp. nov.

(Pl. IX. fig. 1.)

SP. CH. Corona *small, circular;* foot *fully thrice the length of the body, much wrinkled, extensile;* ventral antenna *a single, simple tubercle; a pair of dorsal* hooks *below the corona, adnate at the base;* tube *very short, or absent.*

PLATE C.

Showing the **Coronæ**, **Trophi**, *and* **Foot** *of some of the principal Families of the ROTIFERA.*[1]

Order I. RHIZOTA.

Flosculariadæ	Fig. I. Floscularia campanulata.
„	„ Stephanoceros Eichhornii.
Melicertadæ	Fig. II. Melicerta ringens.

Order II. BDELLOIDA.

Philodinadæ .	Fig. III. Rotifer citrinus.
Adinetadæ .	Fig. IV. Adineta vaga.

Order III. PLOIMA.

Sub-order Il-LORICATA.

Asplanchnadæ	. . . Fig. V. Asplanchna Ebbesbornii.
Synchætadæ	. . . „ Synchæta mordax.
Hydatinadæ	. . . „ Hydatina senta.

Sub-order Loricata.

Division I.

Euchlanidæ .	. . Fig. VI. Euchlanis deflexa.

Division II.

Pterodinadæ .	Fig. VI. Pterodina patina.
Brachionidæ .	„ Brachionus urceolaris.

Order IV. SCIRTOPODA.

Pedalionidæ .	Fig. VII. Pedalion mirum.

[1] See pages 30, 31, 32.

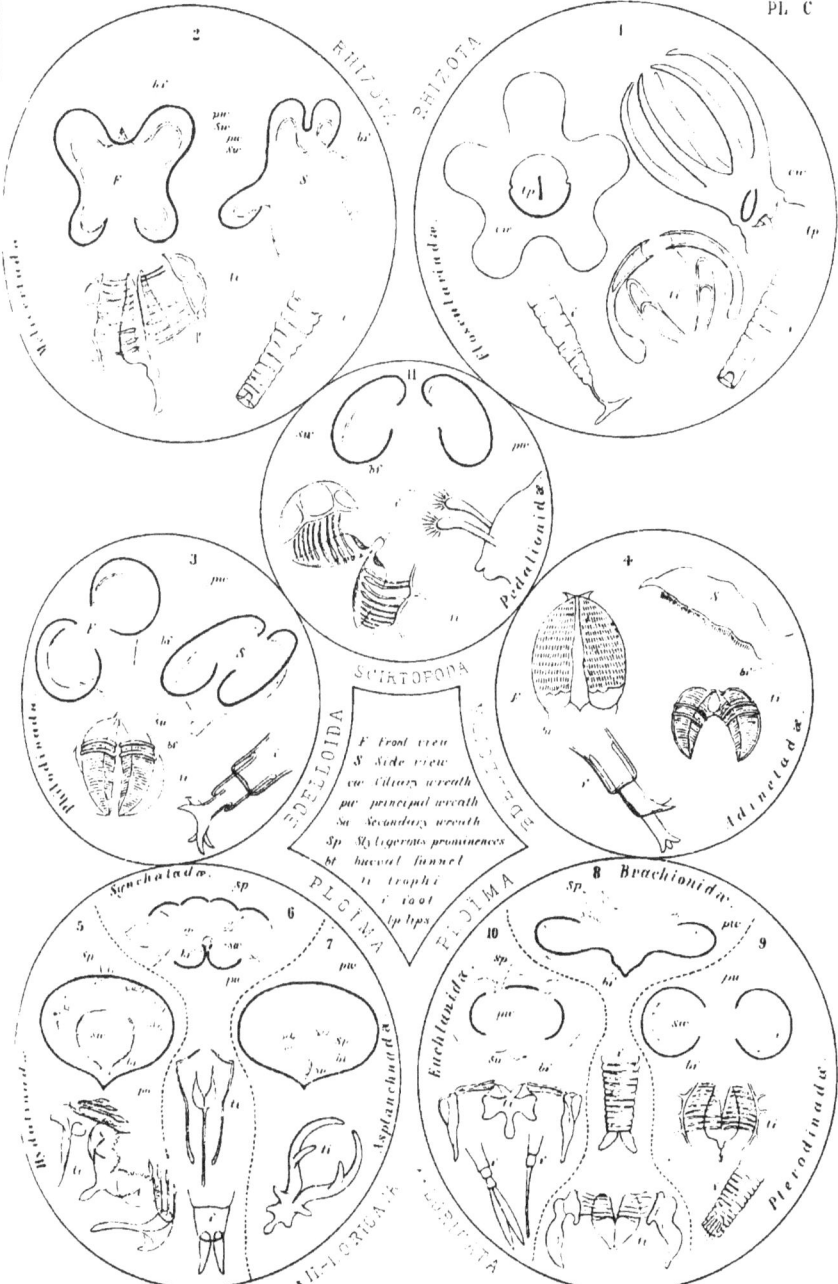

Coronæ, Trophi, & Feet, of some of the principal families of the
ROTIFERA

PLATE D.

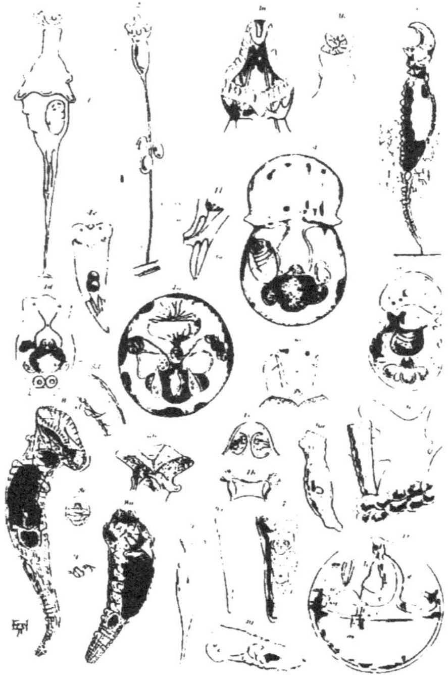

PLATE VI.

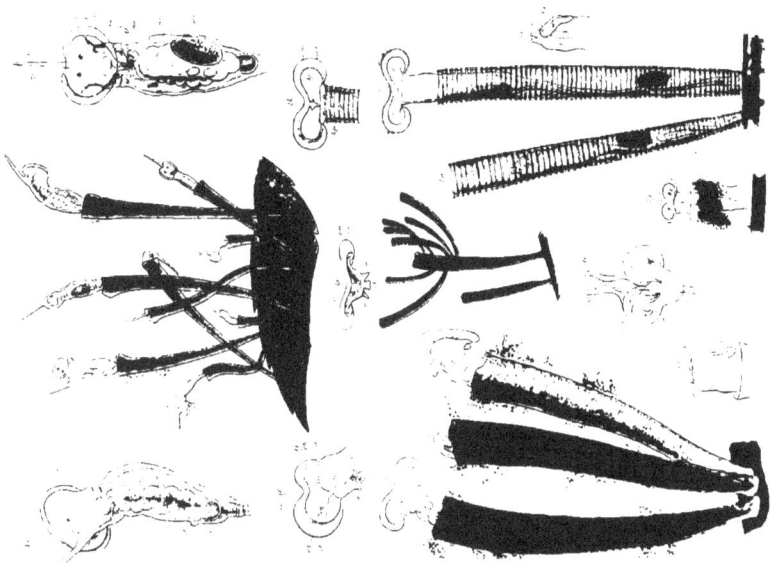

PLATE VII.

PLATE VIII.

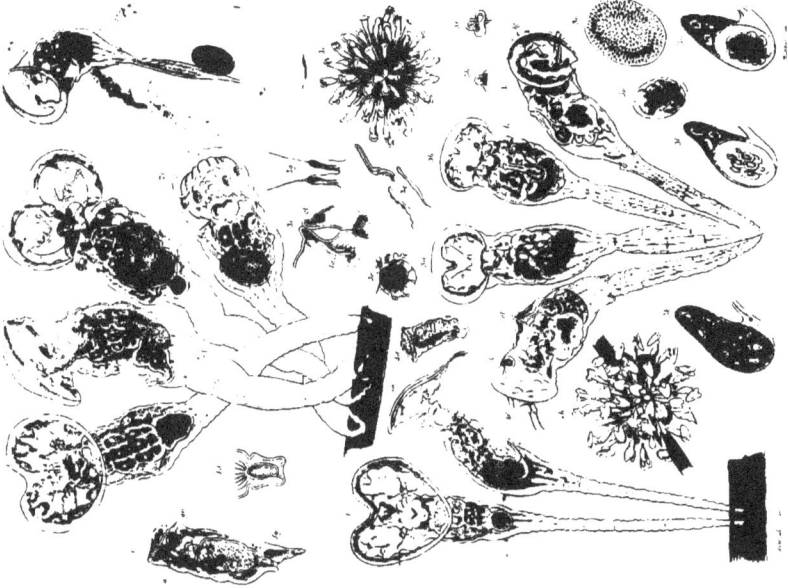

PLATE IX.

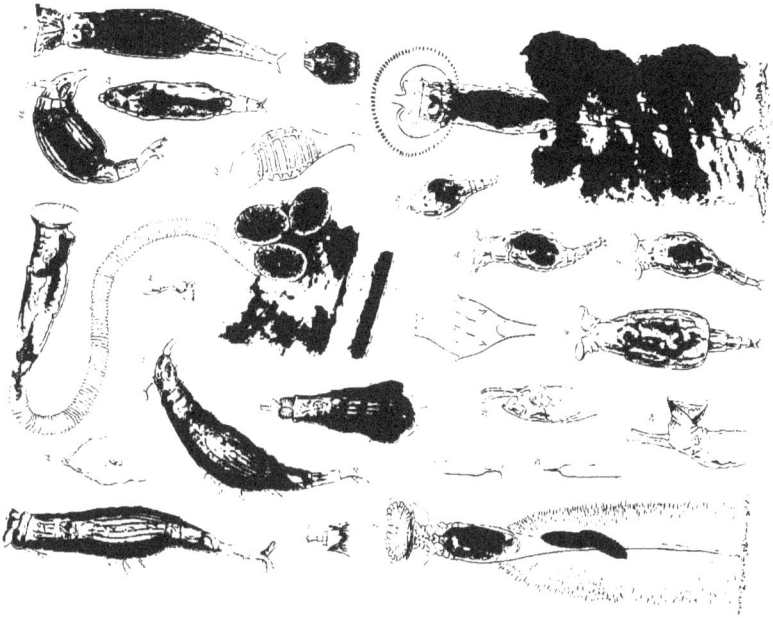

PLATE X.

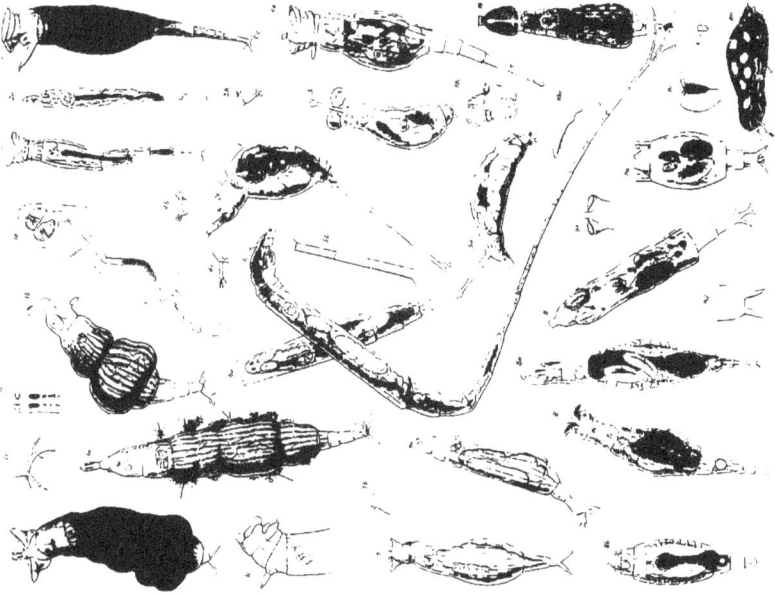

[This very striking species I found on leaves of *Anacharis alsinastrum*, which had been growing for months in a glass jar in my study-window.

The **body** and corona do not vary much from those of other species. In the act of expanding, the summit of the head becomes sub-conical, and is seen to terminate in two small hooks having a common stem, which remind the observer of the protruding head of *Melicerta ringens* (fig. 1*b*). When unfolded, the **corona** does not much exceed the greatest diameter of the body : it appears to form a complete uninterrupted circle. A small round knob on a conical eminence constitutes a **ventral antenna** ; a little below which is seen the mastax ; and, about as much below this, but on the opposite side to the antenna (the dorsal side), the **cloaca** is bounded by a minute wart. The body is of the usual proportions, but the **foot** is of enormous length, being about thrice as long as the trunk and head. It is rather stout (about half the greatest diameter of the body), and of uniform thickness throughout ; thrown for its entire length into transverse close-set wrinkles. These were not obliterated nor perceptibly diminished by the greatest extension that I witnessed ; so that, if this corrugation is a provision for indefinite elongation at pleasure, as one must suppose, then the foot would seem capable of stretching to a length more than ten times as great as I have represented ! Yet I have seen it on repeated occasions contracted in an instant to a condition in which it was not more than half the length of the body, or one-sixth of its former length.

The investing **tube** is reduced to extreme insignificance. The one that I saw would not hold half the body, even if it had no foot. It was invisible, save for a few irregular opaque masses here and there, and for two or three eggs adhering to the margin. As if indifferent to concealment, the foot was attached to a point not near the bottom of even this short dwelling.

The **trunk** is transparent, but tinged with umber-brown. Here and there, within, were multitudes of very small air- (or oil-) globules agglomerated into long masses, which looked like patches of blue-black hue, and had a curious appearance. When I first saw the specimen, two **eggs** were already laid, and presently a third was added, but not under my actual gaze. The eggs were all of a clear yellow, minutely granular.

The **manners** of this creature are as odd as its figure. It is not by any means intolerant of exposure : though sensitive, shutting up and violently contracting on a very slight shock or jar, yet in a moment it is again stretched to its length, and quickly has its corona expanded. The foot is endowed with an extreme flexibility and muscular power, for the animal is constantly (I will not say *swung*, but) shot, from side to side ; just as the body of a snake or of a writhing worm is jerked about in tortuous evolutions.

While I was examining and delineating this example, another appeared, in all essential points agreeing with it, but quite destitute of any apology for a case, the foot being unattached to any object ; the creature being loose in free water, but lying on the same leaf of *Anacharis*. The **male** has not yet been observed.—P.II.G.]

This species in many points resembles Ehrenberg's *Ptygura Melicerta*. The two agree in the small corona, cylindrical body, dorsal hooks, single ventral antenna, many-toothed jaws, and inconspicuous or absent tube. But the prodigious length and extensibility of the great ringed foot of *Œ. serpentinus*, and the animal's extraordinary actions, none of which are mentioned by Ehrenberg, show it to be a different creature. If Ehrenberg had seen only a solitary specimen for a few minutes, in a contracted state, the two might have been supposed to be the same ; but he distinctly says that he had met with many examples. Besides, Herr Eckstein has also found *Ptygura Melicerta* more than once ; and, although he has not seen the expanded corona, his description and figure in all other respects agree with Ehrenberg's.

From the descriptions given by Ehrenberg and Herr Eckstein I am inclined to think that *Ptygura* is an *Œcistes* ; but it is impossible to determine this till the corona has been thoroughly studied.

Length. Not recorded. **Habitat.** On a leaf of *Anacharis alsinastrum* in a freshwater aquarium (P.II.G.).

(Œ. LONGICORNIS, *Davis.*

(Pl. VII. fig. 6.)

Œcistes longicornis　　　　Davis, *Trans. Roy. Micr. Soc.* vol. xv. 1867, p. 13, with figs.

SP. CII. Antennæ *very long and recurved ; tube floccose, irregular.*

Mr. H. Davis found this well-marked species, in abundance, in ponds at Leytonstone, Essex. It is the smallest of the tube-makers. The tube is very irregular and variable in shape ; but, in most of the examples which I have met with, small and tubular at the bottom, while wide and unsymmetrical towards the top. On supplying it with carmine, Mr. Davis saw that the fine coloured particles accumulated in the hollow under the chin, and that they were then rubbed off by the Rotiferon, and left on the top of its case. He thus obtained tubes with crimson tops at least one-fourth of the length of the whole, showing how the structure was gradually formed. The long antennæ, curved back from the ventral surface, and set wide apart, give this Œcistes a very striking appearance.

Length. About ¹⁄₃₀ inch. **Habitat.** Leytonstone, Essex (H. Davis) ; Abbot's Pond, Clifton (C.T.H.) ; Woolston, Cheltenham (P.II.G) ; marsh pools, Fife and Perth (J.H., P.II.G.) : partially distributed.

Œ. PILULA, *Wills.*

(Pl. VII. fig. 2.)

Melicerta, variety No. 2 .	Tatem, *Jour. Quekett Micr. Club,* vol. i. 1868, p. 121, pl. vii. figs. 3, 4.
Melicerta socialis (?)	Collins, *Science Gossip,* No. 85, 1872, p. 9, with fig.
Melicerta pilula (?) .	Cubitt, *Mon. Micr. J.* vol. viii. 1872, p. 5, pl. xxiv. figs. 2–4.
Œcistes pilula .	Wills, *Midland Naturalist,* vol. i. 1878, pp. 302, 317, pl. v. figs. 3, 4.

SP. CII. Antennæ *long ; tube formed of fæcal pellets.*

The first certain notice of this Rotifer is by Mr. Tatem (*loc. cit.*). He gave two excellent and characteristic drawings of it ; saying merely that it was a two-lobed variety of *Melicerta ringens*, without a ciliated cup, and inhabiting a " gelatinous sheath " with adherent fæcal pellets. Its broadly oval [1] corona, however, and the minuteness of the dorsal gap in the ciliary wreath, clearly place it in the genus Œcistes.

As in *Melicerta Janus*, the intestine is large and densely ciliated ; and nearly always contains an oval pellet in the course of construction. Mr. Wills (*loc. cit.*) describes how the animal deposits its completed pellet. He says that it is ejected between the body and the tube, and then caught by the lower margin of the corona. Here it is retained for a few seconds, as if the creature wished to make sure of a proper hold, and it is then, by a sudden retracting of the body, dabbed down on the margin of the tube. The pellets are deposited at irregular times, and the majority of them are so ejected as to be whirled quite away by the coronal currents. Those that form the tube are laid in transverse rings round the body so as to cut the ring obliquely (fig. 2d). The tubes are generally neat, gradually widening to the top, but I have met with some in which the pellets seemed to have been deposited in a most irregular fashion (fig. 2a). These tubes, however, had probably suffered from some accident. The eyes are visible in the half-grown animal.

Length. Scotch specimens up to ¹⁄₇₅ inch. **Habitat.** Sandhurst, Berks (Collins) ; Sutton Park, Birmingham (A. W. Wills) ; Snaresbrook, the tube of unusually large pellets (P.II.G.) : marsh-pools, Fife and Perth (J.H.) : not common.

---

[1] Mr. Cubitt's and Dr. Collins's drawings make the corona so like that of a *Limnias* that I am by no means sure that they are describing Œcistes pilula.

(Œ. BRACHIATUS, *Hudson*, sp. nov.

(Pl. IX. fig. 2.)

[SP. CII. **Corona** *distinctly two-lobed;* ventral antennæ *as long as the transverse diameter of the corona, mobile, non-retractile;* tube *gelatinous, quite transparent, but for adhering matters, subcylindrical.*

The front in retraction is rounded, ending in two geminate hooked points. The ventral antennæ rise from swollen bases wide apart, and diverge on each side nearly in the line of the body. They move independently of each other, often quickly, but not spasmodically. In each can be seen a globose corpuscle within the tip, bearing one seta, whence a double line (probably a tube or a nerve-cord of sensation) can be traced throughout. A mastax of normal form, a long œsophagus, an ample stomach, and a distinct intestine, are plain; from this last a long up-turned rectum leads to the cloaca, at half body-length. Laid eggs, of very long-ovate form, are usually seen, from one to four in number, in the middle part of the tube.

Below the viscera the body suddenly becomes hyaline, and gradually tapers to a slender **foot**, twice as long as the body, through which run many pairs of muscle-threads. At the bottom, already much attenuated, it *abruptly* contracts to an excessively fine thread, which adheres by a sucking disk to the base. The investing **tube is wide and high**, but is absolutely invisible; and can be inferred only from a crowd of minute diatoms (almost wholly of one slender kind) that are entangled in its substance. The cavity of the tube can be easily traced, of equal width throughout, a width determined by that of the body, which, however, rarely retracts sufficiently to enter it. The foot generally displays few transverse wrinkles.

It is not reluctant to display its discal beauties. The **corona** is that of a *Limnias*, composed clearly of two circles united; the dorsal gap wider than the ventral. Each half is conspicuously marked with a thick rib (muscle or vessel?) originating in the common centre, and divaricating, with many branches, nearly to the circumference. But, just within the margin, a concentric band connects these branches, forming a sub-marginal edge, thick and dark, which is constantly thrown into varying puckers by contraction; the whole contributing greatly to the beauty of the corona.

This species, which is among the finest of Rotifera, was discovered in 1882 by Mr. John Hood of Dundee, to whose successful researches this work bears grateful witness. He obtained it from several lochs around; and sent a specimen (which did not survive the journey) to Dr. Hudson, with many notes and sketches. Recently he has favoured me with many specimens, which have freely increased in captivity with me, even in very small phials, so as to be swarming, by scores, for months after their transmission. Thus I have had abundant facilities for study and delineation of the species. It has always occurred attached to filamentous weeds much crowded with impalpable algæ.

Length, $\frac{1}{10}$ inch to $\frac{1}{15}$ inch. **Habitat.** Weeds in Scottish lochs (J.H., P.H.G.): not rare.—P.H.G.]

Œ. (?) VELATUS, *Gosse*.

(Pl. D. fig. 8.)

Megalotrocha velata   Gosse, *Ann. Nat. Hist.* 2 Ser. vol. viii. 1851, p. 198.
Melicerta ptygura (?)   „   *Popular Sci. Rev.* vol. i. 1862, p. 490, pl. xxvi. fig. *d*.

[SP. CII. **Corona** *very large, circular, crossed by thick diverging ribs;* teeth *four in each ramus;* eyes *two, cervical, permanent;* tube *habitually wanting.*

This very beautiful form is aberrant, if indeed it is an Œcistes. My attempts to assign it, first to *Megalotrocha* and then to *Ptygura*, must both be given up. The

H 2

absence of frontal hooks, and the high position of the cloaca, forbid the latter identification. The form of the trophi ; the *very distinct*, small, three-sided intestine (a mark by which the species may usually be identified at a glance), and the absence of any tube-proper, make its position in *Œcistes* doubtful. Dr. Collins, indeed, who has met with it often, and has sketched it in his Note Book, speaks of its occasionally throwing off irregular mucus ; but this seems scarcely a tube. I have lately had many specimens, and have never seen an approach to a tube. The presence of two **eyes**, small but conspicuous, and apparently permanent, is noteworthy. They are situated in the neck, moderately wide apart. Dr. Collins, in one example, figures and describes them as rather close together and near the edge of the corona, which is unaccountable.

The **corona** is a hyaline disk, of beautiful rotundity and of great size, being twice as wide as the body when well expanded ;—shallowly funnel-shaped, but sometimes taking the form of a lovely glass salver, with slightly raised edge, around which the great ciliary waves play nobly. Turbid matters are sometimes poured off through the minute dorsal notch.

The **rectum** may easily be traced, upward from the bottom of the intestine, to the cloaca far above its summit. I have seen an egg, and also faeces, discharged, after which the cloaca protrudes lip-like (8*d*). Around the outside of the corona, investing it to some height, is sometimes seen what seems a vascular tissue of granular yellowish hue (8*b*), apparently movable, whence the specific name. A very good view of the **mastax** enabled me to see that each hemispheric ramus is crossed by four teeth, nearly parallel, whose points project beyond the edge (8*c*, 8*f*).

I found this fine species near London in 1849, in two examples; but never saw it again for many years. Within a few months, however, Mr. Hood sent me water from a pond near Dundee, very rich in forms of Rhizota, among which many specimens of *Œ. velatus* occur. I am tempted to give my ardent correspondent's experience in the procuring of these. It was in the very height of the great snow of early March, in the present year 1886. " I went to the pond to-day to search for *velatus*—a difficult and hazardous task. The pond is more than twelve feet deep, covered with thick ice and snow. As I knew the exact spot . . . I cleared away the snow, and bored a number of small holes in a circle of eighteen inches in diameter ; then thrust down the central plate, which gave me a large hole. I put down my line with sinker and grapnel, but fished a long time with no result. At last a bit of the *Myriophyllum* came up, to which I hope you will find specimens attached ; "—as, I am happy to add, I did.

Length, $\frac{1}{115}$ to $\frac{1}{75}$ inch. **Habitat**. The Black Sea, Wandsworth (P.H.G.); Sandhurst (Dr. Collins); Dundee (P.H.G.; J.H.): rare.—P.H.G.]

Œ. UMBELLA, *Hudson*.

(Pl. VII. fig. 4.)

Œcistes umbella .	Hudson, J. Roy. Micr. Soc. (1878), vol. ii. 1879, p. 1, pl. i.
Œcistes longipes .	Wills, Midland Naturalist, vol. i. 1878, p. 317, pl. v. figs. 1, 2.

SP. CII. Corona *large, nearly circular, crossed with thick ribs ; ventral* **antennæ** *long ;* **tube** *loose, very irregular, clay-coloured.*

Mr. F. Oxley sent me, in June 1878, several specimens of this large and handsome species, which he had just discovered in a pond at Snaresbrook, on the leaves and in the axils of a *Sphagnum*. Its **corona** is so strengthened by thick ribs as to look somewhat like an umbrella. Two of these ribs are very broad, and run across from the ventral to the dorsal side, and when the corona is viewed sideways are seen to project a little above its surface. There is also on either side a branched rib with a triangular space within it like a gusset. When the Rotiferon closes its corona it brings the ribs together, the thinner portions being olded up between them. The thus infolded corona forms a rectangular bundle that is characteristic of the species. The foot is unusually long and slender. Two red eyes are distinctly visible in the adult, just below the surface of the corona.

Length, $\frac{1}{18}$ inch. **Habitat.** Ponds at Snaresbrook, Essex (F. Oxley); and Sutton Park, Birmingham (A. W. Wills); Woolston, Cheltenham (P.H.G.); marsh pools, Perth and Fife (J.H.): rare.

Œ. STYGIS, *Gosse*, sp. nov.

(Pl. IX. fig. 3.)

SP. CII. **Corona** *circular;* **dorsal gap** *not perceptible;* **ventral antennæ** *mere tubercules. Of large dimensions.*

[I found this fine species on September 18, 1885, among impalpable floccose vegetation sent me by Mr. Hood from the Black Loch, near Dundee; a habitat which has suggested a fanciful appellation, and which it shares with its neighbour and rival Œ. *brachiatus.* It equals that species in size, being about $\frac{1}{40}$ inch in height when expanded; its beautiful clear corona, of exquisite rotundity, being $1\frac{1}{50}$ inch in diameter.

The **corona** is strengthened with thick divergent ribs, very visible when viewed obliquely or nearly laterally; but, in a direct front view, these disappear, and the hyaline disk looks like a circular plate of clear glass, marked by a few almost invisible curves. There are no **antennæ** proper; a pair of minute knobs marking their place. But for this, it might be mistaken for Œ. *umbella.* I saw no trace of **eyes.** The body, of the usual form, contracts to a long and slender **foot**; which, at its pedal extremity, dilates into a minute cubical knob, which adheres to a little heap of floccose. A tube of loose gelatinous matter, of brownish hue, surrounds the body, of such a thickness that its diameter is about twice that of the corona. It is undefined to the eye, but carries an agglomeration of extraneous matters entangled in its substance.—P.H.G.]

Length, $\frac{1}{40}$ inch. **Habitat.** Black Loch, Dundee (P.H.G.): rare.

Genus LACINULARIA, *Schweigger.*

GEN. CII. **Cluster** *fixed, of many individuals, with adhering gelatinous tubes;* **corona** *heart-shaped, oblique, with its longer axis placed dorso-ventrally, and a deep ventral sinus;* **dorsal gap** *in the ciliary wreath very minute;* **trunk** *without opaque warts;* **dorsal** *and* **ventral antennæ** *absent.*[1]

L. SOCIALIS, *Ehrenberg.*

(Pl. VIII. fig. 1.)

Lacinularia socialis	.	Ehrenberg, *Die Infus.* 1838, p. 403, Taf. xliv. fig. 4.
,, ,,	.	Leydig, *Sieb. u. Köll. Zeits.* Bd. iii. 1851, p. 452, Taf. xvii.
,, ,,	.	Huxley, *Trans. Micr. Soc.* vol. i. N.S. 1853, p. 1, pl. i. pl. ii. figs. 20-25.
,, ,,	. .	Pritchard, *Infusoria,* 1861, p. 671, pl. xxxvii. figs. 19-25.
Megalotrocha socialis	. .	Gosse, *Popular Sci. Rev.* vol. i. 1862, p. 494.
Lacinularia socialis	. .	Cubitt, *Mon. Micr. J.* vol. viii. 1872, p. 9, pl. xxiii. figs. 3, 4.
,, ,,	. .	Hudson, *Mon. Micr. J.* vol. xiii. 1875, p. 45, pl. xci. fig. 8.
,, ,,	. .	Bartsch, *Rot. Hungariæ,* 1877, p. 19, Táb. i. fig. 1.

This common and very beautiful Rotiferon, the only known species of the genus, is found adhering in clusters, like little balls of grey jelly, to the stems and leaflets of pond-weeds.

[1] Possibly they may be very minute setigerous pimples, which have escaped observation.

The cluster is a lovely microscopic object for a low power with dark-field illumination, and can be easily placed in a live-trough without injury; so that during the hundred and thirty years which have elapsed since its discovery, it has naturally attracted the attention of many observers, and has been the subject of special essays by Professor Huxley and Dr. Leydig. It has, in consequence, been very thoroughly examined, and only a few points of its structure remain as yet doubtful. These, as might have been expected, are in the vascular and nervous systems; and these alone require notice, as the rest of its **internal organisation**, as well as the general plan of the **corona, ciliary wreath, chin** and **buccal funnel,** follows very closely that of *M. ringens.*

The **Vascular System.**—Professor Huxley describes the two lateral canals as arising from a common origin upon the dorsal side of the intestine (rectum), but Dr. Leydig says that the lateral canals start from a common branch opening into a contractile vesicle, which discharges itself into the cloaca : it will be seen that a similar doubt exists concerning the termination of the canals in *Conochilus volvox,* and further investigation is, I think, wanted to make the matter clear.

The **Nervous System.**—Professor Huxley (*loc. cit.*) describes and figures a ciliated cup beneath the chin, just as in *M. ringens* ; and below this cup, underneath the surface, on the ventral side, a " bi-lobed homogeneous mass resembling in appearance the ganglion of *Brachionus.*" This organ he supposes to be the true nervous ganglion. Dr. Leydig, on the other hand, points out two nucleated polar cells, giving off threads, just below the mastax, and two similar ones at the junction of the foot and trunk. These four are, in his opinion, the nerve centres of *L. socialis.* Now, each of these suppositions is liable to the same objection, viz. that it places the nervous ganglion in a unique position. All the known nervous ganglia in the Rotifera lie towards the dorsal surface, and the great majority are near the mastax, and not far from the eyes ; there are no other examples of nervous ganglia on the ventral surface, or in the foot. Besides, in the only one of the *Melicertadæ* (viz. *Conochilus*) in which there is an obvious nervous ganglion, it lies in the normal position ; and M. Joliet describes the nervous ganglion of *M. ringens* as being above the mastax towards the dorsal surface. No antennæ have as yet been observed, and the eyes are visible only in the young.

The **Male.**—I had the good fortune to discover and study the male in November 1874, and published (*loc. cit.*) an account of it, and several other males, soon afterwards. Fig. 1c gives a side view of it. It has a conical head fringed with a wreath of long cilia, a cylindrical soft trunk, and a short, pointed, ciliated foot. There are strong longitudinal **muscles** for withdrawing the corona into the trunk, and several transverse muscular bands in the integument. The **nutritive system** is wholly absent. Two secreting **foot glands** are present, as well as the **lateral canals** and their **vibratile tags.** I several times thought I caught sight of the edge of the **contractile vesicle** behind the upper part of the sperm-sac. There is a large **nervous ganglion** sending threads to a dorsal **antenna** and two red **eyes.** A large **sperm-sac** fills nearly the whole trunk, and ends in a broad tubular, ciliated, and protusile **penis.**

Length. About $\frac{1}{12}$ inch ; diameter of large cluster, about $\frac{1}{4}$ inch. **Habitat.** On water-weeds : common.

Genus MEGALOTROCHA, *Ehrenberg.*

GEN. CH. **Cluster** *fixed, of many individuals without tubes ;* corona *kidney-shaped, oblique, with its shorter axis placed dorso-ventrally, and with a deep ventral sinus ;* dorsal gap *in the ciliary wreath very minute ;* trunk *with four opaque warts in a row on the breast ;* dorsal *and* ventral antennæ *absent.*[1]

---

[1] Possibly they may be very minute setigerous pimples, which have escaped observation.

M. ALBOFLAVICANS., *Ehrenberg.*

(Pl. VIII. fig. 2.)

Megalotrocha alboflavicans .	Ehrenberg, *Die Infus.* 1838, p. 397, Taf. xliv. fig. 3.
,, ,, .	Pritchard, *Infusoria,* 1861, p. 665, pl. xxxii. figs. 371–378.
,, ,, .	Gosse, *Popular Sci. Rev.* vol. i. 1862, p. 191.
,, ,, .	Weisse, *L'Acad. Imp. St. Petersb.* 7 Sér. t. iv. No. 8, 1862, p. 7, with figs.
Lacinularia alboflavicans . .	Cubitt, *Mon. Micr. J.* vol. viii. 1872, p. 9, pl. xxiii. fig. 5.

This is a comparatively rare animal, though it has been known for upwards of a century. Had it been common it would never have been supposed to be a *Lacinularia* without a tube, for only those who have never seen it could make such a mistake. It differs obviously from *L. socialis* in its shape and habits, as well as in its being ornamented with a necklace of four white opaque knobs, "like a row of pearls," stretching from shoulder to shoulder across the ventral surface. These are so placed, that when the animal furls its corona they border the edge of its then pear shaped body.

The cluster radiates in all directions from a small space on the stem or leaf of a plant, and often consists of both adults and their young. Now and then a young brood will swarm out, as in *Conochilus volvox*; and, when swimming away, circling round each other in search of a resting-place, may easily be mistaken for a young cluster of that species. The ways of the animal are unlike those of *L. socialis.* It is a much more timid creature, and does not expand so freely. When a cluster is undisturbed, first one and then another will contract with a sudden twist, to expand in leisurely fashion at its own pleasure ; but every now and then a panic seizes the whole cluster, and they all rush together into a contracted mass, with a curious circular sweep, as if some violent eddy had struck them.

The creature has a peculiar habit of swelling out at times the surface of the corona, so that it is curved up above the ciliary wreath ; but in other respects the two ciliary wreaths, the chin, and the buccal funnel are similar in plan to those of *L. socialis* ; and so is the whole alimentary tract. The two vesicles, which some take to be **salivary glands,** and others mere horny stays to the buccal funnel and mastax, are yellowish ; and so are the **trophi.** The gastric glands are somewhat three-sided in outline when seen ventrally. The lateral canals and their vibratile tags are obvious (figs. 2 to 2c), but the **contractile vesicle** (if any) has not been seen ; neither has it been determined how the lateral canals end posteriorly. No **nervous ganglion** or **antenna** has as yet been seen, and the two red eyes only in the egg and young.

[Mr. W. G. Cocks, of Dalston, has been very successful in keeping this and other fine Rhizota in a tank. He has favoured me with many valuable observations on the habits and development of this species ; and I am indebted to him for a supply of specimens. I noticed that, in one case, when an egg was discharged in my sight, it separated and fell down among the crowded feet of the cluster ; while, in another, it hung awhile to the cloaca after exclusion. Presently a young one was swimming free, probably hatched in the trough. It was colourless, very transparent ; and swam smoothly, evenly, and swiftly, by the rotation of the cilia on its expanded corona. Then came shooting-by a slender creature (Pl. D. fig. 9a), about $\frac{1}{150}$ inch long, with truncate, apparently ciliate, front, tapering regularly to the hind extremity. In this, though full of a bright granulation, I could not detect any viscera, nor other organs. I conjecture this was the new-born **male** of the same ; as fig. 9 is probably a new-born female.—P.H.G.]

In Dr. Weisse's figure (*loc. cit.*) of the female embryo leaving the egg, it will be noticed that the four opaque warts on the trunk, as well as the eyes and mastax, are already formed.

Length. About $\frac{1}{12}$ inch ; diameter of large cluster, about $\frac{1}{8}$ inch. **Habitat.** On plants in gently running streams, Weybridge, Surrey (W. G. Cocks); Kent (Cubitt, Badcock): rare.

GEN. CII.   *Solitary, free-swimming;* body *a perfect sphere;* buccal orifice *on the spherical surface;* principal wreath *dividing the sphere into two hemispheres, and passing above the buccal orifice;* dorsal gap *in the wreath at the pole opposite to buccal orifice;* secondary wreath *a fragment on the under edge of the buccal orifice;* ventral antennæ *extremely minute;* tube absent.

T. ÆQUATORIALIS, *Semper.*

(Pl. D. fig. 11.)

Trochosphæra æquatorialis   Semper, *Sieb. u. Köll. Zeits.* Bd. xxii. 1872, p. 311, Taf. xxiv., translated in *Mon. Micr. J.* vol. xiv. 1875, p. 237, pls. cxx. to cxxii.

Who can complain here that a rough wrinkled skin, complicated external form, and huddled-up organs have rendered his utmost efforts almost fruitless ? Here is a creature whose outer shape and texture have alike reached the very acme of simplicity ; the one translucent as the clearest water, the other rounded into a perfect sphere : an animal created as it were for the study of its internal structure; its organs all symmetrically spread apart in due array, just as if a skilful demonstrator had teased them out with delicate needles ; no one overlapping another, and all deftly hung to the walls of a hyaline globe which not only upholds them, but also displays them to the utmost possible advantage; for it has a band of cilia girdling its entire circumference, and rolling it through the water, so as to present it in every possible point of view.

Trochosphæra has a perfectly transparent spherical body with a principal ciliary wreath running round what may be called the equator, and marking the common boundary of what Prof. Semper calls the "oral" and "ab-oral" hemispheres. In the former lie nearly all the organs of the body ; for only one nerve-thread and portions of a pair of muscular bands are to be found in the other. At one spot in the equatorial ring of cilia there is a break in the wreath, and at the opposite extremity of the diameter passing through this spot is the buccal orifice, which has a very small secondary wreath fringing its oral or lower side.

The various internal organs are so well displayed in Prof. Semper's figure, that only a few points require notice. The trophi, though of somewhat peculiar shape, are malleo-ramate. The lateral canals end in the cloaca, not in the contractile vesicle ; and this latter, according to Prof. Semper, discharges itself into the intestine. The nervous system can be well made out. From the nervous ganglion (*ng*), which lies close to the mastax (*mx*), five pairs of nerves pass respectively to the ciliary wreath, buccal orifice, lateral antennæ (*a'*), lateral canals (*lc*), and eyes (*e*); while a single nerve thread (*n*) passes to the probable dorsal antenna (*a*).

The male is unknown.

It is obvious that if the aboral hemisphere were pressed flat, and the oral lengthened out into a cone, we should have, in the altered *Trochosphæra*, a Rotiferon somewhat resembling one of the *Melicertadæ*. For it would have a buccal orifice laterally placed ; a principal ciliary wreath surrounding the body, with a gap in it on what would then be the dorsal side ; a portion of a secondary wreath passing round the edge of the buccal orifice ; trophi of a Melicertan type; two minute ventral antennæ, and a single dorsal one. On the other hand, the absence of an attached foot, and of a complete secondary wreath, and the difference of habit, make it no easy matter to say where *Trochosphæra* should be placed ; on the whole, I think it should be put among the *Melicertadæ* near *Megalotrocha* and *Conochilus.*

Length. Diameter of sphere, ₃½ inch. Habitat. Ditches in the rice fields of Zamboanga, in the Philippine Islands (Prof. Semper) : rare.

Genus CONOCHILUS, *Ehrenberg.*

GEN. CH. Cluster *free-swimming, of several individuals, inhabiting coherent gelatinous tubes ;* **corona** *horse-shoe-shaped, transverse ;* **gap in the ciliary wreath** *ventral ;* **buccal orifice** *on the corona, and towards its dorsal side ;* **dorsal antennæ** *absent ;* [1] **ventral antennæ** *obvious.*

Take a clay model of an *Œcistes*, and cut off the head by a transverse section through the neck. Lift up the head, and reverse its position, placing the surface of the corona on the decapitated trunk, so that the entrance to the buccal funnel may point towards the centre of the dorsal surface. There will thus be obtained a rough representation of the relative positions of the trunk, corona, and ciliary wreaths in *Conochilus*. Such a violent alteration in the general plan of the *Melicertadæ* might almost seem to entitle *Conochilus* to a family by itself, but its affinities are so clearly with this group that it may well remain here.

On the surface of the corona,[2] close within its edge, and parallel to it, runs a groove, which is broadest and deepest opposite to the dorsal surface, where it is confluent with the entrance to the buccal funnel. The groove grows both narrower and shallower on each side as it approaches the ventral surface, and ceases just before reaching a ventral gap in the corona.

The buccal funnel, except at its wide entrance, is covered by a sloping roof, formed of the uplifted corona, which here rises into a kind of pent-house, notched at its apex. The principal wreath runs round the outer edge of the groove, and is joined, at each side of the ventral gap, by the secondary wreath. This latter fringes the groove's inner edge ; and on reaching the buccal funnel, bends sharply back, rising up each edge of its walls, till it has reached the notch described above ; so that in *Conochilus*, as in other *Melicertadæ*, the entrance to the buccal orifice lies between the two wreaths, and is bordered by the secondary one.

The two known species differ considerably in their modes of clustering, and in their antennæ : they apparently closely resemble each other in other points, but only one has been really studied, viz. *C. volvox.*[3]

C. **VOLVOX,** *Ehrenberg.*

(Pl. VIII. fig. 8.)

Conochilus volvox . Ehrenberg, *Die Infus.* 1838, p. 398, Taf. xliii. fig. 8.
   "   " . Eichwald, *Dritt. Nacht. z. Infus. Russlands,* 1852, p. 520.
   "   " . Cohn, *Sieb. u. Köll. Zeits.* Bd. xii. 1863, p. 197, with figs.
   "   " . Pritchard, *Infusoria,* 1861, p. 664, pl. xxv. 365-370.
Megalotrocha volvox Gosse, *Popular Sci. Rev.* vol. i. 1862, p. 491, pl. xxvi. figs. e, f.
Conochilus volvox . Davis, *Mon. Micr. J.* vol. xvi. 1876, p. 1, pl. cxliii.
   "   " . Bedwell, *J. Roy. Micr. Soc.* vol. i. 1878, p. 176, pl. xi.
   "   " . Hudson, *J. Roy. Micr. Soc.* vol. ii. 1879, p. 3, pl. ii.
   "   " . Imhof, *Zool. Anz.* No. 147, 1883.

[1] Possibly very minute.
[2] Ehrenberg misunderstood the corona of *C. volvox,* and described it as surrounded with a single wreath of cilia and bearing four papillæ on its surface. He placed the buccal orifice on the ventral side, where the ventral gap is ; and suggested that the four papillæ might be a sort of upper lip to the mouth, the edge of the disk itself being the lower one. Dr. Cohn, in his otherwise admirable paper (*loc. cit.*), draws the buccal orifice on the ventral side, and wrongly places the antennæ between it and the dorsal surface. His conical protuberance over the antennæ is also singularly out of shape and proportion. The corona and antennæ were first correctly described by Mr. Davis (*loc. cit.*), whose observations I have repeatedly verified.
[3] *Strophosphæra ismailoviensis* (Poggenpohl, *N. Mém. Mosc.* t. x. 1876) is, I think, a *Conochilus*; with two short separate antennæ lying between a pair of ventral hooks.

SP. CII. **Cluster** *spherical, consisting of many adults and their young;* **tubes** *so compressed together as to be indistinguishable from each other;* **ventral antennæ** *on the corona, between the buccal orifice and the ventral gap, adnate at the base.*

No microscopic object is more beautiful than this lovely globe of living creatures, each bearing its flashing crown of cilia, its ruby eyes and orange-tinted jaws. Fortunately it is as common as it is beautiful, and is equally at home in the Swiss Lakes, in the Scotch Lochs, and in the pools of Hampstead Heath.

The animals all radiate from a common centre (fig. 3c), the extremities of their feet being close together, though not in actual contact. The action of their ciliary wreaths imparts a slow motion to the globe, which rolls along, rising and falling, and often returning on its course, in a very aimless fashion. The globe is formed by the co-operation of its inmates, each contributing its secretion to the structure. The newly-hatched free-swimming young may be seen circling round each other, with their bodies curved, and the extremities of their feet directed towards a central spot in the group. In this way they will swim off in a swarm, not actually cohering, but keeping all close together. I have not been able to follow the process further; but, no doubt, all soon begin to form some sort of tube, and their united secretions fix the swarm together, and at last combine them into a small globe, to which fresh additions are constantly made. A young globe increases its size also, not only by the growth of the original company, but by the addition to it of its newly-hatched young; which, as they emerge from the egg, squeeze a place for themselves among their elders.

But the process has its limits. After a time the globe is too thickly packed, and a young swarm starts off as already described. The largest globes often separate into two portions, each of which soon rounds itself into a sphere; no doubt they are torn apart by the strain on them produced by the opposing action of the ciliary wreaths in opposite hemispheres.

The **internal structure** resembles that of *Melicerta*, but a few points require notice. The **trophi** [1] are tinted orange-red, and so is the lower end of the buccal funnel, where are the lips which form an entrance to the mastax: these latter resist the action of caustic potash. The **stomach** appears to be divided into two chambers, which lie symmetrically on the right and left sides of the body; while between and below them the intestine is curved abruptly back towards the dorsal surface; so that its long rectum ends in an unusually highly placed **cloaca**.[2] Indeed the whole of the viscera are, as it were, tucked high up into the trunk, leaving its lower end empty of all but the longitudinal **muscles**. Six of these run from the head over the trunk, down the broad, transparent, spindle-shaped foot. Five or six bands of transverse muscles cross the trunk, at somewhat regular intervals, from the neck to where it joins the foot. This latter is generally drawn up a little into the body, at its junction with it, so as to make there a conspicuous fold in the integument. The **vascular system** has no **contractile vesicle**,[3] but its function is performed by the cloaca; which I have often watched dilating and emptying at regular intervals. The lateral canals arise in a plexus on each side of the corona, slope downwards dorsally to a similar plexus in each shoulder, and throw out on their way branches, above and below the nervous ganglion, which appear to run into each other. From each shoulder-plexus the canal is continued, still near the dorsal

---

[1] Judge Bedwell (*loc. cit.*) gives a minute, careful, and original account of the structure and action of the trophi.

[2] Mr. Gosse (*loc. cit.*) described and figured the fæcal pellets, which he supposed to be eggs. His account of them is as follows: — "Their form is very peculiar; it appears to be nearly circular, flattened on one side and convex on the other; there is considerable difference in their size; they are of a pale-yellow hue, marked with several blackish specks."

[3] Dr. Cohn (*loc. cit.*) says that each lateral canal ends in a dilated portion or small contractile bladder, which empties itself into the cloaca, and that the two act alternately. The arrangement, however, seemed to me to be that given in the text. I have drawn in Pl. VIII. fig. 3i what I saw. The figure shows a lateral canal (*lc*) ending in what I supposed to be the cloaca (*cll*), and which dilated and contracted regularly. Unfortunately, I have not been able to procure a specimen of *C. volvox* since I read Dr. Cohn's paper.

surface, down to the cloaca. The **vibratile tags** can be easily seen. There is a conspicuous **nervous ganglion** on the dorsal side, just above the neck, and below the two eyes. These latter are beautiful little hyaline spheres (fig. 3*h*) resting on patches of crimson pigment.[1] The two **antennæ** rise from the corona, on the sloping walls of the buccal funnel between it and the ventral gap (fig. 3*a*, *a*). They are adnate at the base, and each carries a bristle that can be withdrawn within a tubular sheath (fig. 3*g*), as in *Melicerta*. The **ovary** is frequently obscured by a large egg, lying across the body, in which the red eyes, moving cilia, and mastax of the young animal are distinctly visible. I have frequently noticed living **spermatozoa** attached to the outside of the ovary : how they can get there it is not easy to see, unless they can find their way from the cloaca, up the lateral canals, and out of the vibratile tags into the body-cavity.

I have watched the formation of an **ephippial egg** from the first enclosing of a considerable portion of the ovary, through the changes shown in figs. 3*k*, 3*l*, 3*m*, to the ultimate production of an egg (fig. 3*n*), covered with a deep layer of hexagonal cells, and bristling with spines, from each spot where the angles of the hexagons meet. As Mr. Davis has well observed, it is a misnomer to call these " winter " eggs, for they occur in all seasons of the year.

The **male** (Pl. VIII. 3*d* and 3*p*), as usual, consists of little else than a sperm-sac and penis. Its general appearance when swimming is shown by Mr. Davis (*loc. cit.*) and its internal structure has been worked out by Dr. Cohn, one of whose figures is reproduced in Pl. D. fig. 10. Dr. Cohn (*loc. cit.*) says that the nutritive system, from mouth to cloaca. is wholly wanting ; that the vascular system is indistinct, though probably present ; that the whole head is occupied by a great nervous ganglion ; and that there are two eyes, which consist of refracting lenses set in pigment. He also describes, and figures, the spermatozoa (Pl. VIII. fig. 3*q*), which he saw under unusually favourable circumstances ; and noticed their attachment to the outside of an ovum (fig. 3*r*).[2]

Length. Diameter of large cluster, about $\frac{2}{3}$, inch ; length of individual, about $\frac{1}{10}$ inch. Mr. Gosse has counted as many as 70, and Mr. Davis 100, in a single cluster. **Habitat**. Lakes, clear ponds, and pools : common.

C. DOSSUARIUS, *Hudson*, sp. nov.

(Pl. VIII. fig. 4.)

Cephalosiphon dossuarius . . Hudson, *J. Roy. Micr. Soc.* 2 Ser. vol. v. 1885, p. 611, pl. xii. fig. 4.

SP. CII. **Cluster** *unsymmetrical, of one adult and a few of its young ;* **tubes** *distinct ;* **ventral antennæ** *below the corona, long, adnate for nearly their whole length.*

This rare species was discovered by Mr. Bolton in September 1884 near Birmingham. It is remarkable for the size, shape, and position of the **antennæ**, which stand on the arched ventral surface like a rifle-sight on the barrel. The specimens that Mr. Bolton sent me were all solitary, carrying with them, as they swam, their cases with the contained eggs ; but Mr. Bolton tells me that the clusters which he usually met with consisted of one adult and a few young individuals of various sizes. On one occasion, too, he saw an adult with one large egg, and four much smaller eggs in its tube. If these latter were male eggs, and the former a female one, this observation would, I believe, be unique.

I had no opportunity of studying the internal structure of this Rotiferon closely ; but I detected no difference in this respect between this species and *C. volvox*.

Length. My solitary specimen was $\frac{1}{50}$ inch. **Habitat.** Near Birmingham (T.B.).

[1] Dr. Imhoff says (*loc. cit.*) that in the specimens in Lake Zug the pigment is black, and Dr. v. Eichwald (*loc. cit.*) found specimens, in ditch-water at Hapsal, in which the eyes were invisible.

[2] The Rev. Lord S. G. Osborne has described (in a letter to the 'English Mechanic,' March 1, 1878), clusters of *Conochilus volvox* which bear at their centres bundles of fine stick-like diatoms. I am indebted to his Lordship for a cluster, mounted by himself, and containing these needle-shaped bodies. They appear to be of three kinds ; they are colourless, and their distinctive markings (if any) are so obscured by the rotiferous jelly, that it is very difficult to say whether they are diatoms or desmids. Lord Osborne's explanation of their presence in the cluster is no doubt the true one ; namely, that they are drawn in point-downwards, bit-by-bit, at each sharp contraction of *Conochilus* into its ball.

CHAPTER VII.

BDELLOIDA.

Si quæ de animalculis infusoriis dici possunt enarrentur, verbaque et oculorum acies sufficerent, dicendi nullus finis esset. Paucissima magnificentiæ et splendoris Numinis optimi maximi documenta prodere mens humana valet ; in plurimis stupet et obmutescit.—O. F. MÜLLER.

He, who feels contempt
For any living thing, hath faculties
Which he has never used ;
    . . . thought with him
Is in its infancy.

WORDSWORTH.

Order II. **BDELLOIDA.**

Swimming [1] *with their ciliary wreath, and creeping like a leech;* **foot wholly retractile** *within the body, telescopic, ending almost invariably in three toes.* [2]

The Leech-like Creepers form so natural a group of animals that all the classifiers of the Rotifera have placed them by themselves. Ehrenberg, Dr. S. Bartsch, Herr K. Eckstein, have arranged them in the family *Philodinæa*; Dujardin, in the order of the *Rotifères*; and Dr. Leydig, in his un-named second family of Rotifera "with a long, jointed, telescopic, retractile foot." This extremely characteristic foot is to be found only in the two families of this order, the *Philodinadæ* and *Adinetadæ*. The longitudinal muscles, which pass down the foot, end at intervals below each other, so that their contraction draws the lowest part of the foot into that just above it, and this in its turn is drawn into the part above, and so on; until the whole foot can be shut up like a telescope, and withdrawn completely into the trunk.

A special interest attaches to the *BDELLOIDA*. Specimens of various species in both the families have been dried, suffered to lie in that condition for three or four years, and then brought to active life again by being placed in water. [3] I have no space to give the history of this question, and of the controversies that have arisen about it, some of which, indeed, are still as lively as ever; I shall therefore confine myself simply to a relation of facts whose reality may be easily tested, and of the satisfactory explanation of them given by Mr. Davis. [4] If specimens of *Philodina roseola* be placed with a little clear water on a slip of clean glass, and the water be quickly dried up, they will all be killed; no watering will revive them. I have tried this scores of times and never met with a case of recovery. But if the rotifers be placed in a cell that contains a little sand, or moss, then the cell may be dried even *in vacuo* over sulphuric acid; and yet, when water is again added to it, in the majority of cases, some of the Rotifera will be found to be still alive: or the cell with its water, sand &c., and the animals may be gradually heated up to 200° Fahr., and yet some of the creatures will probably recover if, when the cell is cool, fresh water be added: or once more, the cell may be laid aside for several years in utter dustiness, and still, on the addition of a few drops of water, the chances are that, in the course of an hour, a few of the animals will revive. Now the real point is obviously this. If a Philodine can be revivified after having been dried in

[1] [The swimming faculty in this order is very subordinate. We never see a *Philodina* or a *Rotifer* sailing smoothly hither and thither, turning waywardly on its course, and roving about with no apparent aim, like a *Microcodon* or a *Euchlanis*. It will bore through a mass of vegetation, and, on coming to its margin, shoot straight away on a voyage of discovery. But the very first new bit of sediment that it meets arrests it; it instantly creeps into this, and makes this its home for a while: as if its natatory powers were used merely for change of place, as distinguished from actual enjoyment in swimming.—P.H.G.]

[2] All the known British species have three toes. Dr. L. K. Schmarda has described some foreign species with only two toes; but I think it probable that he is mistaken.

[3] Mr. Jabez Hogg (*English Mechanic*, Jan. 16, 1885) says that he has seen rotifers revive "after fifteen years' careful seclusion."

[4] *Mon. Micr. J.* vol. ix, 1873, p. 206.

sand over sulphuric acid, or gradually heated to 200° Fahr., or left to the neglected dust of years, why will it not recover from the effects of quick evaporation, without sand, on a glass slip, in the comparatively moist air of a sitting-room ? It has been suggested that, by burying itself in the sand, the animal obtains a covering to protect its internal fluids. But this explanation does not meet the case of Rotifera heated in sand up to 200° Fahr. Surely hot sand at this temperature would be a poor protection for the natural juices of a soft-bodied Rotiferon. On taking some earth or sand containing dried-up Philodines, we shall see them dotted here and there, adhering to the earth's particles, and looking like little red, orange, or white balls. If one of them be picked out, and a drop of water placed on it, after a quarter of an hour or so, a part of the infolded foot will usually shoot out with a jerk, and the foot itself will then gradually lengthen joint by joint. Often, at this stage, the jaws will be seen to be at work, and the head will be driven out with its corona and wreaths still furled : at last these in their turn open, and the recovered animal begins to roam about, or to work for food. Now if, before we moisten the rotiferous earth, a bright light be thrown down on the ball-like Rotifera, it will be seen that each globe has a nearly smooth glittering surface, as if it were coated with a gelatine that filled up the hollows between the stout ridges which run from head to foot. This is the key to the puzzle; for Mr. Davis suggests that the Philodine survives the air pump, oven, and sun-baked gutter, by drawing-in its head and foot into a ball, and then secreting round itself a gelatinous coating which hardens in air and effectually preserves its internal fluids from evaporation. That the gelatinous coat does preserve these fluids Mr. Davis proved by crushing the little balls and finding them all moist within.

But why can the creature do this when sand or moss is present, but fail to do it under much less severe trials in their absence ? Here, too, Mr. Davis's explanation is complete. The water dries more slowly when there is sand or moss in it.[1] The Philo-dines (who are gutter-lovers) are trained in being dried up under these conditions. They naturally creep to the little heaps of sand &c., where the water lingers longest, and, finding it going, contract themselves into a ball, pour out a viscid secretion over their bodies, and prepare for the worst. But all this takes time, and, on the clean glass slip, not only does the water evaporate too quickly, but the animal is too restless in the unusual conditions in which it finds itself to attempt its ordinary defence. As I have often seen, they roam about, vainly seeking shelter, till it is too late; they are overtaken by the rapidity of the evaporation, and dry up never to recover.

[1] It has been questioned whether the presence of sand in a cell, or in a gutter, does retard the evaporation of the water. An experiment or two would soon satisfy a doubter; and a little considera-tion will show the cause of the retardation. When a drop of water is enclosed by three or four morsels of silex, nearly in contact, it is protected by the silex from evaporation everywhere except at the surface. In fact it is in a similar condition to water in an uncorked bottle. Of course the water will dry up in such a bottle at last, but it will evaporate very much more slowly than it would do were it poured out on a marble slab.

If Philodines be left in a zoophyte trough, they will often be found dried up in one of its corners, for the water lingers longest in the angle formed by the bottom and sides. Their instinct leads them too, when other defences fail, to gather together in clusters so as to protect the evaporating water by their own nearly-touching bodies. The Rev. E. J. Holloway, taking advantage of this habit, has suc-ceeded in drying up groups of P. roseola, on slips of clean paper, quite free from sand or rubbish of any kind. He kindly sent me two or three of these slips; and an inspection of them under the microscope confirmed the correctness of Mr. Davis's theory. The fibres of the paper had evi-dently delayed the evaporation long enough to enable the rotifers to resort to their customary defence. Each Philodine is the centre of a patch of glutinous 'secretion, which meets the similar patches, surrounding its neighbours, in a succession of straight lines; so that the whole group has quite a tesselated appearance. Here and there, where fibres pass over or through a group, long tongues of the secretion stretch from the animals to the fibres; and, in one case, a rotifer, that had tried to squeeze itself under some interlacing fibres, had been caught; and, so held, had been moored to half-a-dozen others by radiating viscous cords. The efficiency of the protection was at once shown by dipping one of the slips into water, and watching the Philodines revive as the secretion dissolved.

All the Rotifera seem to possess the power of secreting a viscous fluid, which they put to various uses. The *Rhizota* form their cases of it; the *Ploïma* and *Scirtopoda* draw it out in long threads from spots to which they have adhered, and thus moor themselves to external bodies; while the *Bdelloida*, by coating themselves all over with it, not only resist the extremities of heat and drought, but set at defiance Old Time himself.

Family III. PHILODINADÆ.

Corona *a pair of circular lobes transversely placed;* **ciliary wreath** *a marginal continuous curve, bent on itself at the dorsal surface so as to encircle the corona twice, with the buccal orifice between its upper and lower curves, and having also two gaps, the one dorsal between its points of flexure, and the other ventral in the upper curve opposite to the buccal orifice;* **trophi** *ramate.*

The genera of this family resemble each other so closely that it has often been suggested that they should be reduced to one. They differ from each other chiefly in the number and position of their eyes. One genus has them at the free end of the cylindrical frontal column which forms the anterior portion of the head; and within which, through a fold on its ventral side, the rotatory apparatus can be withdrawn. In another genus they are placed within the neck; while in a third they are entirely wanting. Now, strong objections have been made, by Dujardin and others, to Ehrenberg's use of coloured spots for the purposes of classification; mainly on the ground that it is not certain that they are really organs of sight: and it is true that, in many cases, there is little else to be seen in these so-called eyes but an irregular spot of pigment. On the other hand, some of the Rotifera have unmistakable eyes, consisting of a spherical lens seated on a sort of red, black, or purple cushion. Nerve-threads too, in some species, can be traced from the ganglion to the eyes; and by this means the general positions and appearance of coloured spots, that really are eyes, have been established. Moreover (as might have been expected), the undoubted eyes prove to be in these cases good generic characters. When, therefore, pigment spots are visible in positions that the undoubted eyes usually hold, it is reasonable to consider them to be organs of vision however humble, and to make use of them with other characters in classification. In the genus *Rotifer,* nerve-threads may be seen passing from the ganglion to the eyes in the frontal column; and in *Rotifer vulgaris* Dr. Otto Zacharias has observed "that each of the two carmine-red eye-spots is furnished with a crystalline body." Again, in the genus *Philodina* the position of the red spots with respect to the nervous ganglion is precisely that which is held by such unmistakeable eyes as those of *Conochilus volvox.* I have decided, therefore, to retain the old genera with only a few alterations.

Genus PHILODINA, *Ehrenberg.*

GEN. CII. Eyes *two, cervical.*

The Rotifera comprised in the genus *Philodina*, though technically separated from the rest of the family by their having two eyes in the neck, can be generally recognized at a glance by their greater stoutness of build, by their larger heads, by their more powerful wreaths, and by their habit of so contracting the foot as to form an abrupt division from the trunk. The corona and ciliary wreath would closely resemble those of *Limnias*, were it not for the break in the latter just opposite to the buccal orifice, by which the upper wreath is converted into two segments of circles. The animal, too, holds itself differently from the *Rhizota* while it is feeding; for it slightly arches the dorsal surface so as to throw forward its dorsal antenna; while the *Rhizota* (with the exception of *Cephalosiphon*) reverse this, and arch the ventral surface so as to throw forward the two ventral antennæ. A *Philodina* or *Rotifer*, when creeping, shows no

I

external sign of **corona**: the animal tapers to a point in front as well as in rear. It attaches itself by the tip of its head, and then, releasing and shortening its telescopic foot, takes a fresh hold and arches its body like a leech or geometric caterpillar: it then releases the head in its turn, extends its body, and takes hold again. When it wishes to swim, or to feed, the front of the head is arched backwards; and, by the action of the transverse muscles diminishing the visceral cavity and so driving forward the body fluids, the infolded corona is forced out of a puckered slit on the ventral surface, just where the head joins the trunk. As the corona is pushed out, each disk begins to unfold, and its cilia to act; while at the same time the tapering forepart of the head, or **column**, is thrown backward, falling on the neck, between the lobes of the corona, much as the hood of a cloak falls upon a lady's shoulders, when it has been dropped from her head. The **trunk** is frequently marked with longitudinal furrows, which make it difficult to observe the viscera; and in some species it is tinged with reddish yellow, or brown; while the extremities are usually free from colour. The penultimate joint of the telescopic **foot**, in every species but one, carries a pair of soft tapering processes, called **spurs**, which appear to be useful in giving the Rotiferon a firm hold. The method of attaching itself is this: the three soft toes, rendered sticky by a secretion that exudes from them, are first fastened to the object; then by the contraction of the longitudinal muscles the last joint is drawn over the toes, and the penultimate joint over the last; till the penultimate touches all round the object to which the animal is attached, and the spurs also are brought into contact with it. By this means the Philodine is securely fastened so as to be able to resist the action of its ciliary wreaths, or to dart back without letting go.

There are a few points in the internal structure that call for notice. The **mastax** contains two stout rami, whose appearance is best described by Pl. C. III. fig. ti. They are crossed by two or three principal teeth, with sharp points projecting beyond the internal ridges of the rami, and by a multitude of minute ridges parallel to the teeth. There are also faint indications of the three chitinous loops, attached to each ramus, which are conspicuous in the malleo-ramate type.

The **stomach**, intestine and rectum (unlike those of *Rhizota*) are nearly in a straight line, and the cloaca is situated below the junction of the foot and trunk. The walls of the stomach are unusually thick and absorbent; and become tinged with coloured food almost immediately after it is eaten. The stomach, when empty, is reduced to a tube of narrow bore, whose end dilates into a globular intestine: there appears to be a sphincter muscle separating the two.

Gastric glands, though small in size, are usually visible; and foot glands constantly.

The **contractile vesicle** can be readily seen symmetrically placed in front of the intestine, and though it is difficult to make out the lateral canals, it is generally easy to see some of the vibratile tags. A large triangular **nervous ganglion** lies in the neck, its apex pointing forward, and with a red eye on either side of the apex. The dorsal antenna [1] is long, tubular, and setigerous; its terminal part can be withdrawn within the basal, in the same telescopic fashion as that in which the foot is shortened. No ventral antennæ have been recorded.

The **reproductive system** has been but imperfectly made out. The ovary, with its contained germs, is distinctly visible on either side of the stomach, but no oviduct has as yet been detected.

When a germ becomes developed into an ovum, it is seen, as it grows, to be gradually separated by a constriction from the rest of the ovary, and at last appears to drop off into the body-cavity, in which the young is sometimes completely hatched.

Both the living **young** and the egg have been seen to issue from the cloaca, but it

---

[1] The occipital antenna is normally three-jointed, but in some species the third joint is habitually concealed, in others protruded. Yet this, though characteristic, is not invariable. The terminal joint is three-lobed, each lobe carrying a projecting seta.— P. H. G.]

is difficult to suggest how this is managed, as there appears to be no communication between the body-cavity and the cloaca.[1]

No **male** has as yet been observed among any of the *Philodinadæ*.

P. ERYTHROPHTHALMA, *Ehrenberg*.

| *Philodina erythrophthalma* | . | . | Ehrenberg, *Die Infus.* 1838, p. 499, Taf. lxi. fig. 4. |
| " | " | . . | Pritchard, *Infusoria*, 1861, p. 705, pl. xxxviii. fig. 4. |

[SP. CII. **Body** *smooth with a thick bulging collar;* **corona** *ample with a broad shallow sulcus;* **frontal column** *having no proboscis;* **antenna** *decurved, three-jointed;* **eyes** *conspicuous rond-ovate;* **teeth** *two;* **foot** *stout gradual. Animal hyaline, colourless.*

The characters assigned to this and the following species must not, I confess, be pressed with minute exactitude; nor are they all of equal value. Some of the species do not differ very obviously *inter se*. Still, I think, all *are* specifically distinct; and the cumulate character, which, with some thought and care, I have assigned to each, may aid the scient in discriminating forms too easily confounded.

This first species of the genus is one of the most difficult to be diagnosed with precision; though it is of conspicuous size, and of so common occurrence as to fall very early under the notice of the student. It was almost the first of the Rotifera[2] which I essayed to describe and delineate, nearly seven-and-thirty years ago, from specimens obtained in the north suburbs of London.

The most readily observable feature is, that below the **corona**, itself of ample dimensions, there is a thick prominent neck, bounded by sensible constrictions. Thus it seems inseparable from Professor Ehrenberg's *P. collaris;* and, indeed, I shrewdly suspect these to be but one and the same species, in different stages of growth. The gradual, instead of sudden, transition of the trunk into the stout foot, is another character easily noticed. The **proboscis**, which in this family is general at the tip of the frontal column, seems, here, wholly wanting; the truncate tip having only a slight transverse depression. But the point is peculiarly difficult of determination in this species.

The specific name, besides being repulsive from its uncouth aggregation of unpronounceable consonants, is unsuitable, because undistinctive. The possession of red eyes is common, not only to all *Philodinæ*, but almost to all Rotifera.

Length, $\frac{1}{100}$ inch to $\frac{1}{50}$ inch. **Habitat,** weedy pools, widely distributed: common.—P.H.G.]

P. ROSEOLA, *Ehrenberg*.

(Pl. IX. fig. 4.)

| *Philodina roseola* | . | . | Ehrenberg, *Die Infus.* 1838, p. 499, Taf. lxi. fig. 5. |
| " | " | . | Pritchard, *Infusoria*, 1861, p. 705, pl. xxxv., fig. 490. |

[SP. CH. **Body** *smooth, with no constriction nor swelling at the neck;* **corona** *moderate, with a deep square sulcus;* **antenna** *two-jointed, nearly horizontal;* **eyes** *small, oblong, oblique;* **teeth** *two[3];* **foot** *stout, gradual. Translucent, ruddy in hue.—*P.H.G.]

No doubt it was this common and hardy species which Leuwenhoek discovered in

---

[1] See footnote, p. 103, for a suggested explanation of the difficulty.

[2] [It was the second. *Euchlanis dilatata* was the very first.—P.H.G.]

[3] Lord Osborne kindly sent me many beautiful specimens of the trophi of *P. roseola*, which he had mounted both plain and tinted with carmine. They were in every possible position, so as to admit of a thorough study of the jaws. Among these I found several with two teeth only in each ramus, and a few with two in the one and three in the other. I am aware that, when not accurately focussed, this appearance will be sometimes improperly obtained; but I took care, by delicate focussing, to be able to count the pointed ends of the teeth in each case. Dr. Oskar Schmidt has noticed the same thing in *Rotifer vulgaris* (*Archiv f. Naturgeschichte*, xii. Jahrg., 1 Bd. 1846, p. 69, Taf. iii. fig. 4).—C. T. 11.

the form of li..'e pink balls in the dried-up dirt of a house-gutter ; and whose revivification he describes. It is very common in ponds, water-butts, and housetops ; and will bear to be dried up and reanimated, again and again, without injury. It is, too, most prolific in situations that suit it; and these are sometimes odd enough. Lord S. G. Osborne, for instance, found that the dust of two stone vases in his grounds at Blandford was thick with the little pink spheres of *P. roscola*, and with the white ones of *Adineta vaga* (*Callidina vaga*, Davis) ; and he supplied his microscopic friends for years with this rotiferous dust.

[The **body** is sometimes of a delicate flesh tint, often deepening to full rich red in the cellular walls of the ample stomach, but fainter at the head and foot; it is evident that the tint does not depend on the nature of the Rotiferon's food, and it is quite as glowing in the half-grown animal.

The **corona** is large, with a deeply cleft sinus, and the two wheels of the ciliary wreath are remarkably fine. The **frontal column** is large, cylindrical, truncate, and strongly ciliated. The proboscis has a soft decurved hook on its very front (fig. 4b), which is probably a tactile organ of great sensibility. I believe that I have seen it used for grasping the slender stems and filaments, laying hold of one between the hook and the face. In progress through clear water the creature often makes a perceptible snatch forward, as if it caught prey with the organ, though none was visible. Only in accurate profile, and full extension can this be seen. The broad head of the **stomach** embraces the base of the mastax ; and, when the animal is extended, this viscus is so stretched that the middle portions are drawn thin, while the ends are dilated. There is a short round distinct **intestine** ; and the **cloaca** is at the base of the first joint of the foot. Two earlike triangular **gastric glands** are visible, one on either side of the mastax ; and there is a small **contractile vesicle** which contracts about every thirteen seconds : but the colour of the body and its longitudinal folds interfere greatly with the sight of the internal organs. The **eyes** are of a beautiful pale red ; but are invisible by reflected light. Under pressure eight transverse **muscles** have been distinctly seen, as shown in fig. 4d.—P.H.G.]

Length. When extended, about $\frac{1}{50}$; in a dried condition they are globes of $\frac{1}{100}$ to $\frac{1}{200}$ inch in diameter. **Habitat.** Ponds, water-butts, house-gutters : common.

P. CITRINA, *Ehrenberg.*

(Pl. IX. fig 6.)

Philodina citrina .	Ehrenberg, *Die Infus.* 1838, p. 501, Taf. lxi. fig. 8.
" " .	Gosse, *Tenby*, 1856, p. 299, pl. xix.
" " .	Pritchard, *Infusoria*, 1861, p. 705.
" " .	Eckstein, *Sieb. u. Köll. Zeits.* Bd. xxxix. 1883, p. 353, Taf. xxiv. fig. 14.

[SP. CII. **Body** smooth with a distinct constriction below the swelling disk, but no collar ; **corona** wide, with a deep square sulcus ; **antenna** nearly horizontal ; **eyes** minute, oblong, oblique ; **teeth** two ; **foot** moderately slender, abrupt. Transparent, yellow.

There is great resemblance of form between this and the preceding ; but the differences enumerated above, though mostly minute, help to distinguish it. The colour, however, is the main peculiarity. This is normally a rich clear yellow, like a topaz ; yet specimens occur in which the hue is much paler and duller : and the extremities are always colourless. Under reflected light the creature is an object of great beauty. The citron hue becomes positive, and brilliant, separated abruptly from the hyaline extremities ; while the whole animal assumes a sparkling, glittering appearance, reflecting the rays of light from various points, as if it were carved out of a precious stone.

Though there is no turgid neck, as in *P. erythrophthalma*, there is a more marked constriction than in *P. roscola*, the hemisphere which carries the **corona** being more ventricose. and marked more distinctly with oblique ridges. The period of the contractile vesicle is about sixteen seconds.

It is lively and sprightly. It breeds freely in captivity; in a phial it congregates at the very margin of the water; if they are numerous, a glance along the water line with a lens gives a pleasing sight; it reveals a whole array of the tiny creatures hanging head-downward, in the ridge of water produced by the attraction of cohesion,their amber-like bodies of various sizes, and their broad white crown-wheels all in full play. They do certainly appear to have the instinct of companionship, as Ehrenberg has observed of another species.

Length, $\frac{1}{100}$ inch to $\frac{1}{70}$ inch. **Habitat.** Widely scattered, but by no means common. Near London; Widcombe Pond, Bath; Tenby (P.II.G.).—P.II.G.]

P. MEGALOTROCHA, *Ehrenberg.*

(Pl. IX. fig. 7.)

Philodina megalotrocha	. .	Ehrenberg, *Die Infus.* 1838, p. 501, Taf. lxi. fig. 10.
„ „	. .	Pritchard, *Infusoria*, 1861, p. 705.

[SP. CII. **Body** *smooth, with no constriction nor swelling at the neck;* **corona** *very wide, with no conspicuous sulcus;* **antenna** *three-jointed, unisetate, decurved;* **eyes** *large, roundish;* **teeth** *two;* **foot** *slender, abrupt. Hyaline.*

The great width of the expanded **corona**, and the size of the frontal column and antenna, are very noticeable in this species; as also the plumpness of the body, and its sudden attenuation to form the foot, whence its appearance is somewhat like that of *Rotifer macrurus.* The spurs are small and the toes short, thick, and truncately conical. The **mastax** is set more transversely than usual, so as not to be made out in a longitudinal aspect. Two teeth cross each ramus. In certain lights there seem to be three; but a true adjustment with a power of 300 reveals the projecting points, and shows them to be distinctly two. There is a wide sub-cylindrical **stomach** with a tubular passage, followed by a short and somewhat transverse intestine. This latter discharges, through a narrow but distinct rectum, beneath the second telescopic joint of the foot. Traces of small globose **gastric glands** are seen beneath the mastax. A small **contractile vesicle** lies in front of the rectum. No other portion of the **vascular system** has been recorded. The **ovary** is large with many clear nuclei. The **antenna** appears to have only one terminal seta.

Length. About $\frac{1}{100}$ inch. **Habitat.** Ponds, ditches, &c.; not uncommon.—P.II.G.]

P. ACULEATA, *Ehrenberg.*

(Pl. IX. fig. 5.)

Philodina aculeata	.	Ehrenberg, *Die Infus.* 1838, p. 501, Taf. lxi. fig. 9.
„ „	.	Dujardin, *Hist. Nat. Zooph.* 1841, p. 660.
„ „	.	Eckstein, *Sieb. u. Köll. Zeits.* Bd. xxxix. 1883, p. 352, Taf. xxiv. fig. 15.

[SP. CII. **Body** *beset with spines, having no constriction nor swelling at the neck;* **corona** *not so wide as the body;* **antenna** *two, long-jointed, mobile;* **eyes** *large, nearly round;* **teeth** *three;* **foot** *thick, gradual. Dark brown.*

This species is easily recognised by the spines which have given it a name; but I cannot find these appendages nearly so numerous as in Ehrenberg's figures.[1] Nor are they scattered irregularly over the body, but are arranged in rows on the dorsal aspect. The first row consists of three spines; the next two rows have two each, and are placed

---

[1] Ehrenberg draws as many as twenty-seven spines, irregularly placed, on the dorsal surface. Dujardin (apparently following Ehrenberg's description) says that the body is 'tout hérissé d'épines molles.' Herr Eckstein's description and figure, however, exactly tally with those of Mr. Gosse, only the two spines, which in Mr. Gosse's figures (5b, 5d) of the contracted animal point forwards, are drawn pointing backwards in Herr Eckstein's figure of the uncontracted animal. Very likely the direction of the spines is liable to be reversed by the creature's contortions.—C. T. H.

near together about the middle of the back, and the fourth has also two at the bottom of a strong constriction where the body begins to be attenuated: this pair appears to terminate the body (fig. 5*b*) when contracted, the posterior segment being bent up and forward. The upright spines shown in Ehrenberg's figure I have represented in fig. 5*b*. They are situated one on each side, level with the second pair of dorsal spines. Under pressure, and when seen dorsally, the body assumed the appearance of fig. 5*d*; by which the relative positions of the spines is better shown. The **frontal column** is large and cylindrical, resembling that of *R. macrurus*; the tip, which is truncate, but furnished with a little protrusile proboscis, is ciliated, and capable of a rotating vibration. The **wheels** are large, and the sulcus at the chin is deep.

The **antenna** is large and prominent: it is capable of being erected, of projecting horizontally, or of hanging downward. It consists of two joints, the first tapering to the articulation, which is telescopic, the second dilating to the tip, which is distinctly three-lobed. The deep yellowish umber hue of the body, and its close-set longitudinal furrows, effectually impede the discernment of the internal organs; but I could see that there is a capacious **digestive canal**, which attenuates near the fourth row of spines, and terminates in a **cloaca** at the end of the next segment. Near this point is a small **contractile vesicle**. A large oval transparent ovum was seen in the ventral region of the trunk. As it showed the jaws perfectly developed, I presume the species to be viviparous. This species very seldom expands its corona while stationary; it is impatient, restless, perpetually crawling, or pushing about its closed fore-parts in all directions, while attached by the foot. Generally the expansion of the corona is instantly followed by the detachment of the foot, and away the creature shoots head-foremost, and glides rapidly about the live-box until it is suddenly arrested by coming into contact with some object.—P.H.G.]

Length. When extended, about $\frac{1}{50}$ inch. **Habitat.** Ponds near North London, (P.H.G.): rare.

P. TUBERCULATA, *Gosse*, sp. nov.

[SP. CII. *Scarcely distinguishable from* P. aculeata, *but the surface beset with rough tubercles, instead of spines.* **Body** *much fluted longitudinally;* antenna *with a small three-lobed, club-shaped, terminal joint;* **eyes** *very small;* **teeth** *three, thick;* **spurs** *long, slender, slightly sigmoid, acute. Dark brown.*

A species in most particulars resembling *P. aculeata*, with like club-shaped antenna, much corrugated and plicated, and of a pronounced **wood-brown** hue, but having the recurved spines replaced by rough irregular **tubercles**, has occurred in water sent by Mr. Bolton, Birmingham, from Sutton Park. The sediment of this water, of rich golden brown hue, largely consisting of floccose faecal deposits, as I suppose, full of great Desmids and other algae, has proved usually rich in Rotifera. Among them, grubbing sluggishly, was this uncouth *Philodina*; the whole body clogged with sediment. For some time I supposed that it might be *P. aculeata*, the spines modified into tubercles; but I have found *P. aculeata* repeatedly since, with the spines clear and sharp, exactly as I had drawn them from life thirty-five years ago. And I have also since found numerous examples of this tubercled form, from other waters, in no respect differing from the first; so that I have no doubt of its being an undescribed species. It has very close resemblance to *Rotifer tardus*, except that the two eyes are not in the frontal column, but on the level of the mastax. It has also the manners of that species, slothful, wallowing in the gelatinous floccose, reluctant to move, but ready to rotate.

The **frontal column**, when extruded to the utmost, is tipped with vibratile cilia, amidst which a minute proboscis projects, which is double, consisting of two obtuse blade-like clear laminae, side by side, and slightly divergent.

Length, $\frac{1}{50}$ inch. **Habitat.** Sutton Park, Birmingham (P.H.G.): rare.—P.H.G.]

Genus ROTIFER, *Schrank*.

GEN. CH. Eyes *two, within the frontal column.*

At first sight it seems strange that so humble a form as that of *Rotifer* should have succeeded, not only in attracting universal attention, but also in giving its name to the whole class of animals to which it belongs. But there are good reasons for its having done so. The genus is most widely distributed; it has been found in almost every quarter of the globe: apparently, indeed, wherever it has been looked for. It haunts alike the damp moss on the Great Sidelhorn at a height of 8,000 feet, and the swamps and sea-weeds of the Finnish coast; and seems equally at home under the ice in moor-pools on St. Gotthard, and in the irrigating ditches of the gardens at Cairo. It is, too, as prolific as it is common, and breeds in captivity as freely as when at liberty. More-over, the genus *Rotifer* exhibits the wonder of ciliary rotation in its most effective form; for its wreath when in full action looks precisely like a pair of escapement-wheels of a watch, whirling round at great speed, the cogs looking " sometimes like the ancient battlements of a round tower; at others ending in sharp points, and forming a kind of Gothic crown; now bent all the same way like so many hooks, and now with their ends clubbed like a number of little mallets." How a living creature could possess or use such a contrivance was a hopeless riddle to the early observers; though even some of the earliest expressed the opinion that the appearance was an optical illusion.

But the marvels of the genus did not end here. It had been asserted by Leuwenhoek that the creatures might be dried up for months, and yet be restored to life again. This was enough to fire the dullest imagination; and as also a brisk controversy sprang up about the accuracy of this assertion, the charms of a dispute were added to those of a *Rotifer*, and it was no wonder that the genus secured all its honours; and, like the bramble, was raised by circumstances to a position which it little deserved.

The whole structure of the genus is so nearly that of *Philodina*, that it requires but little notice. The **eyes** are placed close together within the column, generally near the extremity, instead of in the neck, as in the case of *Philodina*. Nerve-threads have been traced to them and light-refracting bodies seen in them, in the case of *Rotifer vulgaris*, by Dr. Zacharias. The only portion of the internal organization that has not been satis-factorily made out is the **reproductive system**. Just as in *Philodina*, no oviduct has as yet been discovered; and no one has explained how it is that the living young pass from their apparently free position in the body-cavity into the cloaca, through which they have been seen to issue.[1]

The **male**, too, is equally unknown; a strange fact, when one considers first that the structure of the female reproductive organs, so far as it has been studied, is that of a typical Rotiferon; and secondly, that thousands of specimens of this genus must have been watched by many practised observers, during the century and a half which have elapsed since the animals were discovered.

---

[1] It is possible that the long thread, which is often seen to pass from the posterior end of the ovary towards the cloaca, may really be, not a muscle, as is usually supposed, but the collapsed oviduct ter-minating in the cloaca. Should this be the case, then the ovum, when it drops from the ovary, does not fall into the body-cavity, as has been supposed, but simply stretches out over itself that portion of the delicate investing membrane, which had up to that moment been shrivelled into a mere cord. As the membrane investing the ovary is of extreme tenuity, it is just possible that it has escaped observation when extended, not only over the ovary, but even over the developed young. If this expla-nation prove correct, there is nothing abnormal in the reproductive system of *Rotifer*. Mr. Gosse quite concurs with me in taking this view of the case: a view which had occurred to him before he read my note.

R. VULGARIS, *Schrank.*

(Pl. X. fig. 2.)

Rotifer vulgaris	Ehrenberg, *Die Infus.* 1838, p. 484, Taf. lx. fig. 4.
„ „	Pritchard, *Infusoria*, 1861, p. 703, pl. xxxv. fig. 476–480.
„ „	Claparède, *Ann. Sci. Nat. Zool.* 5 Sér. t. 8, 1867, p. 11, pl. iii. figs. 6, 7, pl. iv. fig. 1.
„ „	Cox, *Mon. Micr. J.* vol. xvii. 1877, p. 301.
„ „	Eckstein, *Sieb. u. Köll. Zeits.* Bd. xxxix. 1883, p. 355, Taf. xxiii. figs. 6–12.
„ „	Zacharias, *Ann. Nat. Hist.* Ser. 5, vol. xv. 1885, p. 125, pl. v. figs. 1–8.

SP. CII. **Body** *white, smooth, gradually tapering to the foot;* **spurs and dorsal antenna** *of moderate length;* **eyes** *round, small;* **teeth** *two.*[1]

The **nutritive system** of this very common species differs in no way from that of *Philodina.* The walls of the stomach are thick and absorbent, and become tinged at once after the reception of coloured food. The **foot gland** and nucleated **gastric glands** are obvious, and the latter, according to Dr. O. Zacharias (*loc. cit.*), are three-lobed on the ventral surface, but confluent on the dorsal. The same observer gives the following account of the **vascular system.** The **contractile vesicle** opens into the dorsally placed cloaca, and a lateral canal with five vibratile tags can be traced, on each side, down to it from the head. " Each tag has the form of a cylindrical beaker seated by its tapering extremity on the excretory vessel. The beaker is open above, and a broad cilium inserted at its bottom projects a little beyond the aperture. . . Under a power of 1,500 diameters the oscillation of the cilia was so violent that the beakers surrounding them were kept in constant tremulous movement." Of the **nervous system** Dr. Zacharias says : " When examined from the dorsal surface, the anterior portion of the body shows a triangular ganglion placed immediately in front of the mastax. . . The anterior angle of the triangle emits two hardly visible branches towards the eye-spots." Dr. Zacharias thinks that nerve-threads also pass to the extremity of the frontal column and to the antenna ; the former of which bears two long tactile setæ as well as a circlet of small cilia, and the latter a tuft of setæ.

In the same interesting paper, from which I have so freely quoted, there is an account of an oval **parasite,** *Trypanococcus rotiferorum,* which Prof. von Stein discovered, and which Dr. O. Zacharias was at first inclined to consider as a peculiar kind of ovum produced by germination from the inner surface of the cuticle. These parasites were attached to the body-wall at either end of the *Rotifer.* Each was an oval hyaline vesicle, at the free pole of which was a globular finely granulated structure capable of amœboid movements. Where the parasite was attached was a small aperture in the body-wall ; and near this, inside the parasite, a clear ciliated cavity. From this cavity an œsophagus is said by Prof. von Stein to go to the opening in the *Rotifer's* body-wall. Dr. O. Zacharias does not seem entirely satisfied that these egg-like structures were specimens of *Trypanococcus rotiferorum,* but as he found them in abundance it is to be hoped that before long the matter will be cleared up.

The male of *R. vulgaris,* as of every other species, is unknown.

Length. When extended, about $\frac{1}{50}$ inch. **Habitat.** Most widely distributed in fresh water : sea-water, Tay Estuary (P.H.G.) : very common.

---

[1] (Since the diagnosis of all the Bdelloids *inter se* is somewhat indistinct at best, every distinction is valuable. And it may be added to that of *this* familiar species (*R. vulg.*) that it is so strongly, closely, and evenly fluted, as to resemble the *Callidina*; that its length and slenderness, in proportion to its thickness, are peculiar; and that in crawling it often elongates the foot to such an extent as to recall (without any exact resemblance) *R. macrurus.*—P.H.G.)

R. TARDUS, *Ehrenberg.*

(Pl. X. fig. 1.)

Rotifer tardus   Ehrenberg, *Die Infus.* 1838, p. 490, Taf. lx. fig. 8.
"   "   Pritchard, *Infusoria*, 1861, p. 701.
"   "   Eckstein, *Sieb. u. Köll. Zeits.* Bd. xxxix. 1883, p. 358,
                         pl. xxiii. fig. 13.

[SP. CH. **Trunk** *dull brown, viscous, with foreign bodies attached, corrugated longitudinally and transversely;* **extremities** *colourless;* **dorsal antenna** *swollen at the top, large;* **spurs** *long;* **eyes** *shaped like long drops, usually broken;* **teeth** *two.*

This is a large sluggish *Rotifer* of clumsy build, fond of groping among floccose sediment, or of getting within the hollow bracts of a moss and of remaining snugly ensconced there for some time. It frequently appears contracted, the constrictions alternating with prominent swellings, like a sack tied in many places, while the body is fluted almost as regularly as an Ionic column; and its whole surface is covered with a viscous secretion to which floccose matter, small Diatoms, &c. attach themselves; sometimes a long stream of the mucus is dragged behind, with extraneous substances adhering. The colour appears to be wholly external, and to depend, in some degree, on the extraneous matters lodged in the viscous coating; for those portions that are constantly introverted are free from surface-colour, though the viscera have still a slight yellow tint. The corona is large and powerful, and the frontal column is cylindrical, truncate with a minute proboscis at the tip, which does not seem sensibly hooked, or even lengthened, and which projects between and over two small disks each carrying a wreath of vibrating cilia. Within the column, and at some distance from its tip, are the eyes,[1] which are usually long and drop-like in shape, and often broken [2]—one eye often more than the other. The body tapers gradually to the foot, the last joint of which ends in the usual three toes, of which the hindmost is the shortest; all these are curved and claw-like, but truncate. The penultimate spurs are much developed. The dark colour of the trunk, and its close-set corrugations, nearly destroy its transparency, so as to make it very difficult to demonstrate the viscera. By pressure, however, on one occasion, the **intestinal canal** was forced out, attached to the mastax, the ligaments of the anal extremity having given way; it appeared then as at fig. 1*d*; a slender tube, permeating a thick cellular mass, forming the stomach. The mastax has two distinctly separated teeth in each ramus. I cannot but think that *R. citrinus* and *R. tardus* of Ehrenberg are one and the same species.

Length. When fully extended, up to $\frac{1}{36}$ inch. **Habitat.** Near London; Snaresbrook; Birmingham; Woolston; Dundee (P.H.G.): not uncommon. P.H.G.].

R. MACROCEROS, *Gosse.*

(Pl. X. fig. 5.)

Rotifer macroceros .   .   .   Gosse, *Ann. Nat. Hist.* 2 Ser. vol. viii. 1851, p. 202.
Rotifer Motacilla .   .   .   Bartsch, *Die Räderth. b. Tübingen*, 1870, p. 48.
"   *Rot. Hungariæ*, 1877, p. 27, iv. Táb. 34 ábra.

[SP. CH. **Body** *hyaline with longitudinal folds;* **corona** *large;* **spurs** *short, stout;* **dorsal antenna** *very long and mobile;* **eyes** *small, round;* **teeth** *two.*

This form, which I discovered in 1850, and described in " Ann. Nat. Hist." September 1851, is indubitably a good distinct species. It has occurred repeatedly of late. The great length of the antenna, being not less than fully one-third of the whole

---

[1] If, as I suspect, the Rotifer which Dr. Leydig describes as *Rotifer citrinus* is really *R. tardus*, then each of the eyes of the latter Rotifer (according to Dr. Leydig) has a crystalline light-refracting body imbedded in the pigment.—P. H. G.

[2] Herr Eckstein (*loc. cit.* Taf. xxiii. fig. 12) has noticed a similar anomaly in *Rotifer vulgaris*, and Ehrenberg has seen another case in *R. macrurus* (*loc. cit.* Taf. lx. fig. 7,).

animal when rotating, is very observable, as are also its stiffness, and yet great mobility, as well as its unusual number of joints. Nor are the actions of this organ less peculiar, for, in a manner of which I have met with no other example in the class, the animal, in the act of protruding, jerks the antenna from side to side as if feeling with it, wags it about rapidly but not vibratingly, and often taps the water, as it were, with it. As soon as the wheels are quite expanded the antenna becomes still. The organ is very slightly fusiform, quite transparent, and has either a tube or a band running throughout its middle, connected apparently with three very short bristles which project from the somewhat enlarged truncate extremity. Perhaps these are very sensitive, and the band may be a nervous thread which conveys impressions to the brain. Indeed, by careful focusing with a high power, the medial bristle (viewed dorsally) is seen to have a sensible diameter, and to be the continuation of the permeating band projected. Within the first joint, about one-third from the base, the part exterior to it can be retracted. There is not the least bending *at these joints* ; the wagging is *solely from the base*. Two small pale-crimson *eyes* are low down in the column, which is normal in form. An ovate body may sometimes be seen so large as almost wholly to occupy the greater moiety of the abdominal cavity, quite transparent and colourless, in the midst of which is a great mastax, not to be distinguished, *even in dimensions*, from that one which is proper to the animal, but motionless.[1] This of course indicates an unborn *young*, and proves this species to be viviparous. Alongside of this embryo lies a large sac, doubtless the stomach, throughout which the action of vibratile cilia lining the interior is clearly visible. The foot, spurs, and toes are of the usual form. These lowest joints are usually shortened ; the animal habitually sitting, when at rest as well as when rotating, in a squat position, so that they are almost, if not quite, concealed, the long antenna always projected. Generally, save when distended either by digesting food or by an advanced embryo, the whole body is marked with lines, which are longitudinal folds of the skin, not greatly interfering with vision. The **corona** is unusually large and the wheels more than usually circular ; the latter are separated dorsally by a wide sulcus, the lower edge of which is a straight horizontal line. The **mallei** are evidently two-toothed.—P.H.G.]

Dr. Bartsch found this species in the Weilheimer pool, near Tübingen, in company with *Floscularia* and *Melicerta*, and published an account of it (*loc. cit.*) in 1870. He describes the great length of the antenna and the creature's curious actions, " stretching its long antenna far forward and moving it up and down as the water-wagtail does its tail " ; and, under the impression that it was a new species, named it *R. Motacilla*. I have met with this species several times in the clear water of Abbot's pond near Clifton. It was always snugly ensconced in a floccose heap on a stem of alga, or in the axil of a water plant ; and its presence was usually first betrayed by its long antenna, which could be seen waggling about some time before the animal itself appeared.

Dr. Bartsch in "Rot. Hungariæ " (*loc. cit.*) figures the gastric glands, stomach, ovary, and contractile vesicle.

Length, ¹⁄₈₀ inch (P.H.G.) **Habitat.** Near London ; Woolston, near Birmingham ; Stormont Loch (P.H.G.) ; Clifton (C.T.H.) : not common.

R. RAPTICUS, *Gosse,* sp. nov.

(Pl. X. fig. 3.)

[SP. CH. **Body** *clear, brown-stained, not strongly plicate, not enveloped in mucus ;* antenna *long, stout, motionless when extended ;* corona *small.*

[1] I believe that I have seen the distinction between the stomach and the intestine ; and also another great viscus, which must be the ovary. The œsophagus is wide and short. After some hours, the mastax of the embryo worked ; but not rapidly, and only at intervals. P.H.G.

This species I have met with on several occasions in the water from Woolston Pond, though it cannot be considered other than rare. It is one of the larger species, being equal to *R. tardus* in size, with which from its form and colour it may readily be confounded. The colour, however, is a clearer amber, and the comparative absence of the deep longitudinal folds of the skin gives to this species a bright translucency like that of stained glass. It readily expands its **wheels**, which are normal but small. The **antenna**, however, is of unusual dimensions both in thickness and length, equalling *R. macroceros* in this respect. The organ, however, does not wag to and fro, in the curious manner characteristic of that species; nor does the animal squat down on its hinder parts, concealing its foot. The antenna is fusiform, and carries a distinct joint at its extremity, which is, I think, but not tipped with setæ that I could perceive. The thick truncate frontal column bears, near its tip, two large, conspicuous dark-red **eyes**, showing the animal to be a true *Rotifer*. The specific name (from ἅπτομαι, to touch or test) alludes to the presumed function of the prominent antenna.

Length. About ₂⁄₅₀ inch. **Habitat.** Woolston Pond (P.H.G.): rare.—P.H.G.]

R. MACRURUS, *Schrank.*

(Pl. X. fig. 4.)

| *Rotifer macrurus* | . . . | Ehrenberg, *Die Infus.* 1838, p. 490, Taf. lx. fig. 7. |
| " " | . . . | Pritchard, *Infusoria*, 1861, p. 704. |

[SP. CII. **Body** *white, hyaline at the ends, plump, suddenly attenuated to a slender and very long foot;* **corona** *large;* **spurs** *small;* **frontal column** *long, cylindrical, truncate;* **dorsal antenna** *of moderate length;* **eyes** *usually small and round;* **teeth** *two.*

This is a large and well-marked species, imposing and attractive. Its stout corona, large wheels, and plump body are much like those of a *Philodina*; it can be generally recognized by the sudden break in outline between the trunk and the foot, by the great length of the latter, and by the long stout column, which stands well up above the expanded wheels. The **spurs** are unusually short for so large a *Rotifer*, being of about the same length as the three toes. The foot consists apparently of eight joints, almost all of unusual length. There is a short, but bulging neck. The distinction between the **stomach** and intestine is often visible, and the rectum extends through (what appears to be) the whole basal joint of the foot. There are two oval **gastric glands**, as well as a pair of club-shaped glands in the foot. A small **contractile vesicle** can be readily seen, and so, with some little difficulty, can the lateral canals and vibratile tags. Dr. Leydig failed to make out these last, but both Dr. Bartsch (*loc. cit.* Tübingen) and myself have seen three tags on each side. The **nervous gangl on** has not been noticed. The shape of the **eyes** appears to be variable. Dr. Leydig says that he met with some specimens in which the pigment spots were hemispheres much cut out in front, and in others were lengthened out into a row of points lying behind one another. The former had light-refracting bodies seated on them, which the latter lacked. A pair of **muscles**, inserted into the neck, pass to the great constriction behind; another pair, inserted in the shoulders, pass down into the first joint of the foot; and a third pair start from near the same point as these last, and pass to points where the first pair end. Each joint of the foot has at least two longitudinal muscles on each side, which pass into the joints above them. Under strong pressure the whole foot is seen to be crossed with transverse muscles of which at least twenty can be counted. High pressure shows numerous muscles in the trunk also. These are in the form of broad bands alternating with spaces of equal width. I saw the birth of a **young one** twenty-four hours after it had (apparently) escaped from an oval membrane into the body-cavity. It passed head-first through the cloaca in a few seconds. It was compressed during the passage, and the parent's cloaca was but little disturbed. This young one extended, before its birth, from the base of its parent's foot to the base of the column ; and, when born, was $\frac{1}{50}$ of an inch long, or about two thirds of its parent's length.

Length. When extended, about $\frac{1}{30}$ inch. **Habitat.** Widely distributed : common.— P.H.G.]

Genus ACTINURUS, *Ehrenberg.*

[GEN. CII. *The whole animal excessively long and slender ; eyes two, frontal ; teeth two, converging.*

Technically, there is little but its extreme length and tenuity to distinguish this genus from *Rotifer ;* the only difference that Ehrenberg gives (viz. the number of the toes) being founded on error, since all the *Rotifers* have three, and not, as he assigns to them, two. Yet he has done well to separate the present form. The diminution of thickness, and the great development of length, impart a peculiar *facies.* which at once precludes the possibility of mistaking it for any species of *Rotifer,* as soon as seen. Even in contraction the trunk is not sensibly thickened, never swelling in the middle as in the other genera of the family it does.—P.H.G.]

A. NEPTUNIUS, *Ehrenberg.*

(Pl. X. fig. 6.)

Actinurus Neptunius .	Ehrenberg, *Die Infus.* 1838, p. 496, Taf. lxi. fig. 1.
„　　„ .	Pritchard, *Infusoria,* 1861, p. 704, pl. xxxv. figs. 181-4.
„　　„ .	Eckstein, *Sieb. u. Köll. Zeits.* Bd. xxxix. 1883, p. 359, Taf. xxiv. fig. 17.

[SP. CII. **Frontal column** *short, carrying the eyes near its base ;* **corona** *small ;* **trunk** *long, slender, cylindrical ;* **foot** *protrusile to twice the length of the trunk ;* **spurs** *small, two-jointed ;* **toes** *very long, recurved.*

From its excessive length and tenuity the appearance of the creature is very remarkable. It may be likened to a cylindrical tube out of which protrude a great number of draw tubes from both extremities, principally the posterior one. The head is peculiar, when viewed laterally ; ovate in form, the frontal column very short, and the eyes being oval, dark and large, with the antennal tube projecting obliquely backwards, it presents a ludicrously strong resemblance to the head of a rabbit. The **rotatory organs** are small and seldom unfolded ; the **eyes** of an intense red, almost black. The eight or nine joints which constitute the **foot** are of extreme slenderness, and its **spurs** consist each of two joints ; the first club-shaped, the second very slender and acute. The three long, slender, cylindrical, diverging **toes,** are flexible, and commonly bent outward. Owing to the slenderness of the body the viscera are greatly elongated. The **mastax** is at a considerable distance from the corona, and is reached by a long buccal funnel. Each **ramus** bears two inwardly-converging teeth. The digestive canal is apparently undivided, and originates directly from the mastax : with, I think, two small **gastric glands.** I think I detected a **contractile vesicle.** The viscera, however, can be demonstrated with difficulty, partly owing to the strong longitudinal corrugations in the dorsal region of the trunk, and partly to the creature's incessant contractions and elongations. The ovary is obvious, and the appearance of the eggs suggests that the animal is viviparous. This singular creature is lively in its motions ; and it is a curious sight to see the immense length of foot suddenly thrust forth from the body, in which it had been completely hidden, the starting out of the horizontal processes, and the diverging of the long toes, as these are successively uncovered.

Length. Fully extended, $\frac{1}{13}$ inch : closed, $\frac{1}{46}$ inch. **Habitat.** North London ; Leamington ; Caversham : rather rare (P.H.G.).—P.H.G.]

Genus CALLIDINA, *Ehrenberg*.

GEN. CII. **Eyes** *absent*.

Although this genus differs technically in so slight a degree from those which we have considered, yet it can be generally recognized by its slender, pointed shape, its smaller size, and the abundant corrugations of its skin. Its internal organization, with the exception of the eyes, is so precisely that of *Rotifer*, that it requires no special description. One of the species (Mr. Gosse's *C. bihamata*) possesses a very obvious pair of those dorsal hooks which are so characteristic of the nearly related *Melicertadæ*, and which are to be found also in the still humbler and more closely connected genus *Adineta*.

C. ELEGANS, *Ehrenberg*.

Callidina elegans	.	.	.	Ehrenberg, *Die Infus.* 1838, p. 482, Taf. lx. fig. 1.
"	"	.	.	.	Perty, *Zur Kenntniss kleinst. Lebensf.* 1852, p. 43.
"	"	.	.	.	Pritchard, *Infusoria*, 1861, p. 702, pl. xxxiv. figs. 470-3.

[SP. CII. **Body** *fusiform, abruptly enlarged centrally, strongly fluted, collared;* **frontal column** *thick, truncate, ciliate, with a decurved proboscis;* **jaws** *with no prominent teeth;* **foot** *thick;* **spurs** *moderate.*

This species, the only one of the genus known to Ehrenberg, I find not uncommon in various waters: but it is only by careful observation that it can be distinguished from its congeners. The **corona** is scarcely wider than the body, the double disk being very little more than a full circle, or two circles *very* slightly separated, when seen quite vertically. The **column** is short, unusually thick, with a minute acute proboscis over-arching the dorsal edge of its tip. The **antenna** longer than width of corona, slender, subequal throughout, flexible, truncate, carrying (apparently) a short terminal brush of fine setæ. A swelling collar above the antenna. The **trunk**, bounded above and below by a strong transverse fold, is abruptly swollen, impressed with strong longitudinal plaits, frequently visible as the animal turns. The **foot** is thick; the penultimate spurs middling; the three toes well developed. The outline is very variable. The trunk is tinged with clear yellow-umber, which is abruptly defined at both ends; the extremities being colourless, and very hyaline when stretched in extension.

The **trophi**, when viewed quite dorsally, have much of the form seen in the *Rhizota*, as figured (for *Limnias*) in my mem. "On Mand. Org." fig. 71 : the rami being long and pointed, and having a projecting handle-like knob. After keen scrutiny, I cannot discover any teeth crossing them, or anything to break the uniformity. I hence conclude that there are only the very close minute lineations, which Ehrenberg describes. A vanishing shadow of a curved line on each side may represent the mallei, but quite undefined. The restlessness of the animal precludes fine definition. It is most impatient, incessantly moving, not still for an instant. It rarely swims, but perpetually crawls by alternate elongation and shortening, in a course excessively devious.

Length, $\frac{1}{16}$ to $\frac{1}{25}$ inch. **Habitat.** Sutton Park Ditch; Woolston Pond: rather common (P.H.G.).—P.H.G.].

C. BIDENS, *Gosse*.

(Pl. X. fig. 8.)

Callidina bidens	.	.	.	Gosse, *Ann. Nat. Hist.* 2 Ser. vol. viii. 1851, p. 202.

[SP. CII. **Surface** *closely corrugated;* **spurs** *minute, conical, pointed;* **teeth** *two.*

I obtained this species at Messrs. Smith and Beck's in 1849, from a nearly putrescent infusion of hay, in which it was swarming almost to the exclusion of everything else. Its manners differ much from those of other *Philodinadæ*. It is, if I may use the term, very wild, shooting about with swiftness in an impatient manner, with a peculiar mingling of swimming and creeping; proceeding in this way all about the live-box by the hour together, so rapidly and irregularly that the motion of the stage can scarcely keep it in the field. It is much bolder than the other members of the family, keeping its wheels in rotation all the time it is attached; and though a sudden jar, or the impact of another animal, will cause it to close them, it is but for an instant. I have never seen it contract on alarm into a short round bulb; far less remain quiescent in such a condition for hours, as *Rotifer* and *Philodina* do. It is spindle-shaped, the central region of the body always having an angular prominence; but this varies its situation. sometimes the upper part, sometimes the middle, sometimes the posterior of the abdomen, projecting, according to the position of the viscera at the moment; the creature thus assumes various candelabra-like forms, as shown in figs. 8, 8a. The head, when extended, terminates in a thick rounded **column** which is ciliated; when the wheels are expanded, the column appears small, square, and truncate, fits in below the wheels, and does not project beyond their surface. The **foot** is moderate in length; the **spurs** of the penultimate joint are very minute cones, and the last joint has one small stiff point behind, and two soft cylindrical protrusile lateral toes, truncate at their extremities. The whole surface of the body is covered with minute irregular and close-set corrugations. The buccal funnel is very long, and the **rami**, which are very small, are each crossed with two teeth. A scarcely appreciable œsophagus leads to an enormous and very mobile **stomach**: in one specimen this organ appeared to be composed of a number of spherical cells; in others of a minutely granulated texture: it ends in a short rectum.

There are two corrugated **glands** in the foot, and a **contractile vesicle** (whose period is forty seconds) with the usual lateral canals. A long **ovary** with double rows of rudimentary ova occupies each side of the stomach, and two large eggs are commonly seen, of a bright pellucid appearance, but sometimes dark and granulate. The whole animal is crystalline, and usually colourless, but I have seen a specimen in which the wheels were of a delicate pale citron-colour, and another in which the intestine was of nearly the same cinnabar hue as in *Philodina roseola*, though not so brilliant. It is noteworthy, seeing that the creature is eyeless, that the specimens which I first possessed had been kept in the dark; expressly because " it was found speedily to die, if kept in the light." With the phial of water I obtained, I impregnated two vessels, one of which I placed in a window, the other in a dark corner. Five months elapsed, when, on examination, the species was abundant in the darkened phial: but in that in the window I could not find a single specimen.

Length. About $\frac{1}{15}$ inch. **Habitat.** Pools on Hampstead Heath; lake in Kew Gardens (P.H.G.).—P.H.G.]

C. PARASITICA, *Giglioli.*

(Pl. X. fig. 9.)

Callidina parasitica	Giglioli, *Quart. J. Micr. Sci.* N. Ser. vol. iii. 1863, p. 237, pl. xi.

SP. CII. **Spurs** *stout, conical, as long as width of the contracted joint;* **teeth** *two. Parasitic on the limbs of crustacea.*

This species was discovered by Mr. H. Giglioli as an epizoic parasite on the thoracic and abdominal appendages of *Gammarus pulex* and *Asellus vulgaris*; and was figured and elaborately described by him *loc. cit.* According to Mr. Giglioli the body is very transparent and colourless, fusiform in shape, and without the angular prominence at its central region which is so distinct in *C. bidens.* The **corona** consists of two small circlets of short cilia, and is rarely expanded; the animal usually contenting itself with crawling like a leech over its host. There is a distinct **alimentary canal** surrounded by a yellowish-green cellular mass, a broad pyriform ciliated stomach, narrowing gradually to a bent intestine, and again widening into a broad and richly ciliated cloaca. No salivary or gastric glands have been seen. There is a large irregular contractile vesicle, with a period of about thirty seconds, and two very small lateral canals ; the vibratile tags, however, have not been made out. The dorsal antenna is large, and divided into three lobes at its tip, but no setæ were visible. No eyes have been seen either in the adult or young. [I found several examples in June 1885, on the thoracic limbs of *Gammarus pulex*. To Mr. Giglioli's interesting details I have little to add. He says that out of 700 or 800 *Gammari*, he had not found one free from these *Callidinæ*. My experience is not quite confirmatory of this constancy. Out of four *Gammari*, I found *Callidinæ* on only one. They adhered to its limbs merely as other Bdelloids do to any surface. The " suckers," mentioned by Mr. Giglioli, are no organs of special function, connected with parasitism, but are the three truncate toes common to the whole order.

My examples, four in number, crawled off their nurse presently, on the subjection of the latter to pressure in the live-box, and moved actively about in the free water ; contracting, and elongating, and readily swimming. Their hue was a pale straw-yellow, becoming clear ochre-yellow in contraction, which deepened to umber in the middle of the alimentary canal, and in the maturing ovum : but of a glassy translucency. The pair of spurs at the bottom of the penultimate joint of the telescopic foot, are stout, thick, long and pointed. It is difficult with creatures so extremely variable in outline to give any measurements which are not too vague to be distinctive : yet, as compared with these organs in *C. bidens*, I may say that in *C. parasitica* they are as long as the diameter of the penultimate joint itself, when thickened to its utmost by retraction ; whereas in *C. bidens* their length does not equal half the width of the joint in the same condition. Their bases are separated by a horizontal space equal to their own breadth (fig. 9d). The joint itself is closely and minutely fluted.—P.H.G.].

Length. Up to ⅓₀ inch. Habitat. Parasitic on *Gammarus pulex* and *Asellus vulgaris* (Mr. H. Giglioli ; P.H.G.).

C. BIDAMATA, *Gosse*, sp. nov.

(Pl. X. fig. 7.)

[SP. CII. Frontal column *bearing two hooks, mutually crossed.*

This species I found in the sediment of a phial of water that had been standing on my study table for more than a month, originally sent to me by Mr. H. Davis, dipped by him from a pool near Snaresbrook in June 1885.

It has not any obvious peculiarity to distinguish it from its congeners, except that the column is terminated by a pair of acute hooks, set on the same plane, and crossing each other transversely, like the blades of a pair of shears. These at first sight suggested the *C. vaga* of the friend to whose kindness I had been indebted for this stranger. A moment's observation showed that it was not that species, now elevated by Dr. Hudson to the rank of a genus, *Adineta*. Yet the peculiar structure in question may well be considered as a marked approach to it. Minute hooks, terminating the column, are, indeed, common to all the species of the *Bdelloida*, but usually soft, obtuse, decurved, and single. The whole trunk, somewhat swollen, is strongly scored with longitudinal folds of the skin ; a dozen or more in number. The double corona was readily expanded, and the animal glided freely and swiftly through the free water. It is small, as in *Callidina* generally : the column with its terminal hooks was projected (or rather not retracted) during the coronal rotation. The dorsal antenna is placed unusually far back (see fig. 7a) ; it is small, obtuse, oblique, connected with a dorsal tubercle ; not seated on it, but, so to speak, leaning on its front slope. It is scarcely in advance of the mastax, when this is in its normal position. I did not see on it either cilia or setæ.

Length. Estimated at about ₇₀⅒ inch. Habitat. Pool near Snaresbrook (P.H.G.).— P.H.G.]

Family IV. ADINETADÆ.

Corona *a flat, prone surface;* **ciliary wreath** *the furred ventral surface of the corona;* **trophi** *ramate;* **frontal column** *soldered to dorsal surface, and ending in two hooks.*

The family *Adinetadæ* has been formed to receive one genus, which itself contains at present only one species. It has been separated from the *Philodinadæ* on account of its lacking the usual corona of two circlets, and of its having in lieu of it a mere furring of a flattened, ventrally placed, portion of the head, which in some degree resembles the face of the genus *Proales.* In other respects the organization is that of the *Callidina,* except that the frontal column, which is so striking a feature in the other *Philodinadæ,* and which is tossed aside like a hood when the coronal wheels are expanded, is here soldered as it were to the dorsal surface, and projects slightly beyond it, bearing two curved hooks.

Genus ADINETA, *Hudson.*

GEN. CII. Eyes *absent.*

A. VAGA, *Davis.*

(Pl. X. fig. 10.)

Callidina vaga . . . . Davis, *Mon. Micr. J.* vol. ix. 1873, p. 201, pl. xiv.

SP. CII. Body *smooth, colourless, with longitudinal corrugations;* **spurs** *short, finely pointed;* **teeth** *two.*

This species was discovered by Mr. H. Davis in 1867, along with abundant specimens of *P. roscola,* in a parcel of pink dust sent to him by the Rev. Lord S. G. Osborne, and found in some open stone vases in Lord Osborne's grounds at Blandford. These vases, at times, become partly filled with rain, and the wind drives into them dead leaves and other matters, which by their decomposition seem to afford suitable food.

The front of the **head** is somewhat like a thumb in outline, and on its dorsal surface is what appears to be the trace of a **column** soldered to the head, projecting slightly beyond it, and bearing two transverse hooks. The **corona,** though nothing but the flat ventral surface of the head, yet shows a sort of division into two parts, owing to the absence of cilia on a broad median line leading to the buccal funnel : the rest of the surface is densely furred with minute cilia. The base of the corona, just at the animal's neck, rises on either side above its general plane, and forms a well-marked ridge. Mr. Davis says that each ridge is strongly serrated, and draws them with teeth pointing forwards. I could see no such serrations, though the strong cilia, that here lead to the entrance of the buccal funnel, frequently produced a fleeting appearance of serration : but all my specimens were small and young, and possibly the serrations were not yet developed. The pathway (if it may be so termed), through the coronal cilia, leads straight to a long buccal funnel imbedded in a thick fleshy mass, and ending at the mastax, each ramus of which bears two main teeth and a multitude of fine parallel striæ.[1] When *Adineta* is feeding, it curves the flat corona so as to bring its two furred halves opposite to each other, and at the same time draws together the transverse ridges at its base. In this way a ciliated semi-cylinder is formed as a prolongation of the buccal funnel, and minute atoms may be seen rapidly streaming down the tract clear of cilia, into the funnel's entrance. The stomachs of my specimens were all filled with clear yellowish particles, but of what substance I could not make out. I could see no salivary or gastric glands, but the **foot-glands** were conspicuous. The **contractile**

---

[1] Mr. Davis measured these, and found there were from thirty to forty thousand in the inch; in my specimens the striæ were about fifty thousand to the inch.

vesicle was also plainly visible : Mr. Davis says that its period is about twenty seconds in the young, and from two minutes to five minutes in the adult. By compression I obtained an unusually distinct view of the lateral canals (fig. 10*b*), which showed clear against the grey ground of the ovary. Mr. Davis has seen five vibratile tags on each side, but not the lateral canals ; and I succeeded in finding two pairs of tags attached to the canals : one pair just below the transverse ridges of the corona's base, and another mid-way between this point and the contractile vesicle. There is a **dorsal** antenna ; but no **eyes** ; the **nervous ganglion** has not been made out.

A. raga may be recognised at once by its peculiar movements, which are unlike those of any other Rotiferon I am acquainted with, and are also extremely embarrassing to the observer. It fastens itself by its foot, extends its body to its full length, and then swiftly contracts itself. Nor is this all ; for, instantly extending itself again, it alters its direction, as well as its distance from the surface to which it is attached. In this way it will work around its toes, as around a pivot, compelling the observer to make incessant alterations of the stage and focus. In fact very little is to be learned of its structure, while it is permitted to go free. When imprisoned in a very closely flattened drop of water, it is a little more manageable, for it ceases then to dart backwards and forwards ; but on the other hand it makes up for this by constantly gliding about at a very fair pace. It is quite incapable of swimming in the open, and may now and then be seen rolling ignominiously from the top of the glass cell to the bottom ; but it glides with ease over the surfaces of stems, plants, or glass, by the reaction of those surfaces on the ciliated head. It is, if possible, a hardier creature even than *Philodina roseola* ; for Lord Osborne tells me that he has repeatedly found that, in a mixed gathering of the two, *A. raga* will survive successive dryings and moistenings which have proved fatal to all the former.[1]

Length. From $\frac{1}{90}$ to $\frac{1}{30}$ inch. **Habitat.** In some stone garden-vases at Blandford (Lord S. G. Osborne) ; in a mill-pond at Petit Bot, Guernsey (C.T.H.) ; frequently in dried gatherings of *Philodina roseola.*

---

[1] Mr. Davis's experiments (*loc. cit.*) show that this is not always the case.

CHAPTER VIII.

PLOÏMA

(IL-LORICATA).

Curiosum nobis ingenium Natura dedit, et artis sibi ac pulchritudinis suae conscia spectatores nos tantis rerum spectaculis genuit; peritura fructum sui, si tam magna, tam clara, tam subtiliter ducta, tam nitida, et non uno genere formata solitudini ostenderet.— SENECA.

For Beauty, Good, and Knowledge are three sisters
That doat upon each other, friends to man,
Living together under the same roof,
And never can be sundered without tears.
TENNYSON.

CHAPTER VIII.

Order III. PLOÏMA.

Swimming with their ciliary wreath, and (in some cases) creeping with their toes.

It has been already seen [1] that our typical Rotiferon was drawn from the ranks of the *PLOÏMA;* and rightly so, for the number of its genera, the abundance of its species, the restless energy, perfection of structure, and superior intelligence of its members, clearly entitle the third order to be considered the typical one. It is true that *Pedalion* makes a still nearer approach to the *Arthropoda* in its structure than does any species of the *PLOÏMA*, and must be ranked above them in the scale of the animal kingdom; but it is almost (if not quite) the only representative of its order, and therefore unfitted to be taken as a type of the class.

In the Free-Swimmers the mastax reaches its highest development, and is often used like the mandibles of an insect. Mr. Gosse and I, as well as other observers, have seen these active creatures seize their prey with their jaws, and watched them nibbling the floccose sediment on the stems of water-plants, or slitting up the cells of algæ and the skins of infusoria in order to extract their contents. Indeed, the snapping of the protruded jaws among some of the *Notommatadæ* is so vigorous that it is difficult to see it and not to fancy that we hear the snap; and on one occasion, even, the fierce atom has been seen to give itself a fatal lock-jaw by its outrageous snatch.

In this order, too, as might have been expected from their habits, the nervous system is conspicuous, the ganglion being large, and the nervous threads from it to the various organs of sense more easily traced than in the first and second orders. The eyes, too, have often obvious lenses, and the tactile organs are numerous and well-developed. The vascular system, whose probable respiratory functions must be of the utmost importance to these restless animals, is unusually well developed. It spreads a network of coiling tubes close under the cuticle, and not unfrequently ends in a contractile vesicle so large as to fill, when distended, an important portion of the body-cavity.

In fact, the whole structure of the order shows its members to be well equipped for the energetic life which observation proves them to pursue. They haunt the algæ on pond walls, coast along the water-line among decayed leaves and floating *débris*, dive down to the bottom to explore the muddy sediment, or boldly put off from shore, and sail out even into the middle of such a lake as that of Zurich. No doubt the marvellous coronæ of the *Rhizota* and *Bdelloida* will always continue to attract the expert and amateur alike; but to obtain an adequate notion of the structure of the Rotifera, and of what may fairly be termed their mental capacities, the inquirer must turn to the study of the *PLOÏMA*.

Sub-Order IL-LORICATA.

Integument *flexible, not stiffened to an inclosing shell; foot, when present, almost invariably furcate, but not transversely wrinkled; rarely more than feebly telescopic, and partially retractile.*

[1] P. 1.

Family V. MICROCODIDÆ.

Corona *oblique'j transverse, flat, circular;* **buccal orifice** *central;* **ciliary wreath**
*a marginal continuous curve encircling the corona, and two curves of larger cilia, one on
each side of the buccal orifice;* **trophi** *forcipate;* **foot** *stylate.*

Genus MICROCODON, *Ehrenberg.*

GEN. CII. **Eye** *single, centrally placed, just below the corona.*

M. CLAVUS, *Ehrenberg.*

(Pl. XI. fig. 1.)

Microcodon clavus .	Ehrenberg, *Die Infus.* 1838, p. 395, Taf. xliv. fig. 1.
„ „ .	Pritchard, *Infusoria,* 1861, p. 665, pl. xxxii. figs. 371-2.
„ „ .	Grenacher, *Sieb. u. Küll. Zeits.* Bd. xix. 1869, p. 187, Taf. xxxvii. fig. 2.

This curious Rotiferon, for which, though the only species of the genus, it has
been found necessary to make a new family, was discovered by Ehrenberg in 1830. It
has since been described by Dr. Max Perty and Dr. H. Grenacher (*loc. cit.*), and has
been found in England several times by Dr. Collins, at Sandhurst, and lately by Miss
Davies, at Woolston. It has, however, escaped the notice of the majority of observers
during the last fifty years, in some measure no doubt owing to its small size; for
though its whole length is $\frac{1}{125}$ inch, more than one-half of this is taken up by a long
narrow foot, so that the actual body of the animal does not much exceed $\frac{1}{300}$ inch.
Ehrenberg placed it in his family *Megalotrochæa;* but neither in its ciliary wreath, its
trophi, nor its foot, does it resemble the *Rhizota.* The **corona** is a flat circular disk, set
obliquely on the trunk, and with its dorsal edge pointing forwards.[1] A complete ring of
minute cilia edges the disk, and these perform the office of driving the food to the buccal
funnel. The entrance to this latter is near the centre of the corona, a little towards the
ventral surface. It lies between two curves of large unequal cilia, of such lengths and
so arranged that they form on each side an oval border to the buccal orifice. Usually
these large cilia are at rest; and *Microcodon,* under the action of the smaller cilia, either
glides along swiftly or oscillates gently to and fro over the same spot, as if it were moored
by a thread from its single toe. It is possible that this curious hovering over one place
may be due to the mutually opposing action of the minute cilia of the two halves of the
corona, but it always gave me the impression that the animal was at these times moored
by a viscous thread to some spot on the glass. Every now and then, whether gliding
along or hovering, the creature darts suddenly forward with the utmost swiftness, accom-
plishing this by vigorous strokes from the two rows of larger cilia. It would seem that
it has unusual control over this apparatus, for Dr. Grenacher has seen, in an injured
specimen, these inner cilia lifted and depressed one by one; and has traced to them
what he supposes to be nerve-threads arising from the depth of the corona. A further
peculiarity noticed by this observer is that the corona remains expanded, no matter how
the animal be treated.

The **trophi** (fig. 1c) consist mainly of two ribbed rami, attached to a long narrow
plate (the fulcrum), which is seen edgewise in the figure. I think, too, that I detected
delicate pointed unci on each side of the incus. The whole are included in a long conical
mastax, closely resembling that of *Polyarthra,* and pointing downwards towards the

---

[1] Ehrenberg says that its shape is that of a transverse figure 8; but Dr. Grenacher, Dr. Collins,
Mr. Gosse, and myself all agree that it is circular. Mr. Gosse, however, points out that, when the
corona is inclined to the line of sight, it does look somewhat like that of a *Limnias.*

ventral surface. Dr. Grenacher says that there is no separation between the stomach and intestine, but this Mr. Gosse has distinctly seen. The latter says: " When I first detected the animal, the intestine occupying the gibbous swelling of the hind abdomen was clear, save for a considerable well-defined mass of orange-red ; but, on resuming my observations on it an hour or so later, the intestine was not distinguishable from the stomach, the whole being of a deep rich sienna-brown, with oil-globules of various sizes scattered throughout it." The ovary appears to be divided into two distinct portions, and a clear reddish spot, somewhat like an oil-globule, but of unknown function, lies between the stomach and ovary. A **contractile vesicle** is conspicuous above the foot, and Dr. Grenacher notices his having seen indistinctly the lateral canals, but not any vibratile tags. There are two spherical **gastric glands**, and just above these, at the head of the mastax, is a round **nervous ganglion**, on which is seated a splendid **eye**. It is a purple ball, resting on purple plates curving round the ganglion, so as to give the whole a curious likeness to a jockey's cap (fig. 1b). Two of these stripes appeared to have been displaced in my specimen, but I cannot tell if this was an accident or if their position is normal. I could only make out these details by flooding the animal with transmitted light. There are a dorsal and two lateral antennæ, all more setigerous warts ; and the foot bears just above the toe on the dorsal side three bristles, which Dr. Grenacher says are erectile. The same observer has noticed that the longitudinal muscles which move the foot, and are continued down into it from the body, are all striated. The foot is divided into three joints, of which the first and last are small, and it ends in a single toe. It is freely moved from one side to the other, round its basal joint, and is sometimes laid flat to the ventral surface.

The **Male.**—[A female had been playing in my live-box within an area formed by bounding filaments of *Myriophyllum*. Presently I saw a slender worm, about as long as this charming subject itself, of almost aerial transparency, very slender, darting about the same limited area. It was a nearly perfect cylinder, but gradually tapering to an acute extremity, which may possibly have been a minute conical toe. The front, slightly bent downward, was transversely truncate ; its circular margin carrying a wreath of locomotive cilia, by whose vibrations it shot vigorously and rapidly about. The whole body was refractive of light, but one vesicle, situate about two-thirds from the front was more intensely refractive. This I suspect to have been the sperm-sac. I could detect no other organ or viscus in the animal, but the entire length and breadth was full of minute granules. My grounds for suggesting that this was the male *Microcodon* are but inferential. First, the motions were exactly imitative of those which I had just been watching in the female—swift glidings hither and thither, occasionally varied with moments of sudden pausing, and again with still more sudden and invisibly rapid starts and springs to a distance. Secondly, its appearance at the same time, in the same dip, and in the same limited area with the female, which itself is a rarely occurring species with me : and thirdly, the apparent attentions which the supposed male paid to the female, every now and then coming close to her in his devious travels, though only to shoot by her. The area was quite open at one end ; yet for a long time, and not till after many sailings to and fro, did he assert his freedom, when she presently followed.—P.H.G.]

Length. From $\frac{1}{15}$ to $\frac{1}{30}$ inch, of which the foot is more than half. **Habitat.** Sandhurst (Dr. Collins, P.H.G.) ; Woolston Pond, Hants (Miss Davies, P.H.G.).

Family VI. ASPLANCHNADÆ.

Corona *sub-conical, with one or two apices ;* **ciliary wreath** *single, edging the corona ;* intestine and cloaca *absent.*

The *Asplanchnadæ,* though singularly beautiful Rotifera, are yet of a low type of structure, for their stomach is a blind sac, and they reject all fæcal matter through the mouth. The family contains the genera *Asplanchna* and *Sacculus*, which, while separate I from all other Rotifera by the absence of intestine and cloaca, differ also from each other in several important points. The incudate trophi of *Asplanchna* are massive forceps quite free from an inclosing mastax, and capable in consequence of even plunging down into the œsophagus ; but the forcipate trophi of *Sacculus* are feeble hooks and blades, inclosed in a grape-shaped mastax, and admitting of only a slight protrusion from the mouth. The stomachs also of the two genera are widely different : that of *Asplanchna*, a spheroidal bag of moderate dimensions ; that of *Sacculus*, a large sac with six great cæcal appendages nearly filling up the whole of the body-cavity : and, further, *Sacculus* carries its eggs attached by a thread to a hollow of the posterior surface, while *Asplanchna* produces its young alive.

Genus ASPLANCHNA, *Gosse.*

GEN. CH. Corona *with two apices;* **trophi** *incudate, not inclosed within a mastax ;* **stomach** *of moderate size, spheroidal. Viviparous.*

The various species of this genus differ from each other mainly in the presence or absence of the foot, in the shape of the trophi, in the number and colour of the eyes, in the size and arrangement of the vascular system, and in the external shape and appendages of the male.

It was in this genus that the first male Rotiferon was discovered ; and, indeed, the great size and transparency of the female, and its habit of producing its young alive, give unusual facilities for the study of the males. Many of the Rotifera deposit their eggs here and there on the stems and leaves of algæ &c., so that it is impossible to identify the males when hatched ; but in the genus *Asplanchna* the male may be readily seen alive in its mother's ovary. It is a creature of the greatest delicacy, like a bubble of the clearest glass ; and yet the various species can be easily distinguished by the differences of figure. In one the male is a mere reproduction of its parent on a reduced scale ; in another it bears two sleeve-like processes on its sides : in a third, four such processes ; while one has its ventral surface prolonged into a sheath for the penis.

A. EBBESBORNII, *Hudson.*

(Pl. XI. fig. 3.)

Asplanchna Ebbesbornii.   Hudson, *J. Roy. Micr. Soc.* 2 Ser. vol. iii. 1883, [p. 621,] ls. ix. x.

Female *with one dorsal, one ventral, and two lateral* **humps ; eye** *single ;* **rami** *with singly pointed ends, not serrated ;* **contractile vesicle** *expanding to more than half the body-cavity ;* **vibratile tags** *often forty on each side, and arranged in straight lines ;* **ovary** *horseshoe-shaped :* **male** *with two additional, lateral, humps, below the neck.*

This fine and rare Rotiferon was discovered by Mrs. Tupper Carey in 1880, in a duck-pond in the vicarage of Ebbesborne Wake ; and, strange to say, this, at present, is its only known habitat. It differs from all other species of the genus in its outline, which is not bell-shaped, and in its possession of four sleeve-like prolongations of the cuticle ; one on the dorsal surface, one at the hind end of the ventral, and one on each side of the body below the head. All these appendages are empty of organs, and somewhat flaccid when the animal is swimming quietly ; but, when it draws in its head, they are driven out stiffly from the general surface. Fine muscular threads tie their extremities to various parts of the body, and serve to shorten the processes when the head is again protruded : the two side appendages have their tips connected by a fibre passing straight through the body from the one to the other. The **head** is conical with two apices ; and in the hollow between them, a little towards the ventral surface, lies the buccal orifice, with two small style-bearing prominences on either side of it. The **ciliary wreath** is a simple ring of cilia surrounding the coronal cone, and bent down inwards at the ventral surface, to the buccal orifice. This latter opens directly on the jaws, which are two stout hooked rami with their fulcrum plate imbedded in a horse-shoe-shaped cushion : doubtless, as Mr. Gosse suggests, the third lobe of the mastax : there are no other lobes, so the jaws are practically free. The muscular bands which open and shut these formidable nippers are shown in fig. 3c, and can be seen with ease. Immediately below the horseshoe-shaped ring, and partly embraced by it, is a chamber or **pharynx** formed by a delicate membrane strained over four curved rods, which hang downwards, and are joined together at the bottom by cross pieces. This curious contrivance resembles somewhat in shape the silk well of a lady's work-table. By suddenly pulling the four rods apart at the top, the animal causes a partial vacuum, and any prey near the buccal orifice is instantly engulfed. I have seen this happen often, even to Rotifera of considerable size ; for *A. Ebbesbornii*, like all the *Asplanchna*, is an indiscriminate feeder, and will swallow even such awkward mouthfuls as *Triarthra longiseta* and *Brachionus Bakeri* ; the latter of which I have seen with its posterior spines actually piercing the stomach and body walls of the *Asplanchna*, while the former I saw head-downwards in the œsophagus, with its long propellers stretching right up almost to the buccal orifice. From the pharynx, on its ventral side, stretches a very long transparent and extensile œsophagus, down which run ribbon-like muscular threads. Not unfrequently it is loaded with food, so that the creature seems then to have a stomach of twice the usual size. The gastric glands (figs. 3c, 3d) have each a large duct leading to the stomach, and the cells imbedded in them lie together in clusters of four and five. The spherical **stomach** has thick cellular walls with a clear round nucleus in each cell : it is often divided for a time into two unequal portions by a deep muscular constriction, as in fig. 3a. There is not a trace of intestine or cloaca : the hind end of the stomach is merely attached to the body by two fine threads. The fæcal matter is rejected through the mouth ; it is slowly driven up by successive contractions of the stomach and œsophagus, till it reaches the pharynx, where it is grasped by the jaws and tossed out through the buccal orifice. There is a rectangular **nervous ganglion** just under the corona, touching the buccal orifice on its dorsal side, and carrying a crimson eye. A nerve-thread passes from each of the four corners to one of the antennæ. Two of these latter are on the dorsal side of the body about half-way down, and two others are on the coronal apices. There is yet another pair on the corona just above the buccal orifice, and to these branch nerves are sent from those of the second pair. The **ovary** is horseshoe-shaped, flattened at the ends, and studded with germs which are often clusters of cells (fig. 3h). The ova are always developed at the hind-surface of the middle of the ovary : and, when they attain some size, they drop off into the ovisac, a funnel-shaped pocket with its broad base attached to the contractile vesicle. The ovisac ends in an oviduct, which opens on the ventral surface in a transverse slit. Occasionally I have met with specimens that had as many as three or four **ephippial eggs** in the ovisac at once ; but generally there is only one maturing ovum, or a young animal lying across the parent with its head presented to the opening of the oviduct. The birth of the **young** is almost instantaneous ; and I have seen it expelled with such force that its stomach was driven right through its mouth, so that it was born with its ciliary wreath half-way down its body. The ephippial eggs [1] are circular, corrugated, and often tinged yellow ; as is sometimes the ovary itself.

The **male** (fig. 3i) is rather more than half the size of the female, and bears two small additional appendages below the neck. As usual, the entire digestive tract is absent. There are, however, some rounded masses adhering to the dorsal surface, just below the hump. Possibly these may be a kind of stored-up material to compensate for the male's inability to take nourishment.

The **sperm-sac** with its enclosed spermatozoa, and the penis, are shown in fig. 3f. The penis is a tube with delicate longitudinal furrows sheathed in soft granular masses,

---

[1] Two ephippial eggs are shown in fig. 3.

and having a ciliated opening. It can be drawn back by two pairs of muscles attached to the dorsal surface. Short muscular threads help to draw and direct it forward; but its vigorous outward movement is mainly due to the compression of the body-fluids by transverse muscles. The spermatozoa can be distinctly seen in motion in the sperm-sac, and they are of the two forms drawn in fig. 3k. The **vascular system** (fig. 3f) is alike in both sexes, but is much better seen in the male. The flocculent ribbons which support the lateral canals are unusually large and long, and are looped up here and there by threads, and also tied in the same way to the body-walls. They appear, themselves, to be tubes of a loose granular stuff, with clear cells (fig. 3j) imbedded in their walls. Possibly it is through these cells that the perivisceral fluid finds its way into the tubes and thence into the lateral canals. These latter are much smaller tubes, connected with the first, but meandering along their edge; and they have attached to them an amazing number of vibratile tags; often more than forty on either side. The contractile vesicle, to which the flocculent ribbons and lateral canals are obviously attached, swells out in the female, at times, so as to occupy nearly two-thirds of the body. In the male it is smaller, but in both sexes it is covered with a fine muscular network, which is constantly compressing it into ever-varying and graceful shapes. The **muscular system** is best seen in the male, owing to the absence of digestive organs, but is so plainly shown in the drawings that it requires no further explanation.

Length. Nearly $\frac{1}{50}$ inch. **Habitat.** Ebbesborne Wake, Wiltshire (Mrs. Tupper Carey): very rare.

A. Brightwellii, *Gosse.*

(Pl. XII. fig. 1.)

A diœcious rotifer allied to the genus Notommata . .	Brightwell, *Ann. Nat. Hist.* 2 Ser. vol. ii. 1848, p. 153, pl. vi.
An infusory animalcule allied to the genus Notommata .	Dalrymple, *Phil. Trans.* 1849, p. 331, pls. xxxiii. xxxiv.
Asplanchna Brightwellii and A. Bowesii .	Gosse, *Ann. Nat. Hist.* 2 Ser. vol. vi. 1850, p. 23.
Ascomorpha Anglica	Perty, *Zur Kenntniss kleinst. Lebensf.* 1852, p. 39.

SP. CII. **Female** *without humps;* **eye** *single;* **rami** *with doubly pointed ends, not serrated;* **contractile vesicle** *expanding to about one-fourth of body-cavity;* **vibratile tags** *on each side varying from about ten to twenty, and arranged in a straight line;* **ovary** *horseshoe-shaped;* **male** *without humps.*

Mr. Brightwell discovered this species in a small pond immediately without the city of Norwich in 1841. Both sexes were in abundance, so that he was able not only to make out the structure of the male, but also to witness several acts of copulation; thus, for the first time, establishing beyond all question the diœcious character of at least one species of the Rotifera. The **female** differs from that of *A. Ebbesbornii* in the following points. It is bell-shaped, possessing none of those **humps** which are so striking a feature in the former species. The **jaws** (fig. 1b) differ slightly in their proportions and shape, and Mr. Dalrymple (*loc. cit.*) detected on either side of the stout rami delicate curved rods, which no doubt are the unci. The **gastric glands** are kidney-shaped, the **contractile vesicle** somewhat smaller. The **ephippial egg** of this species is also circular, but it has on its outer covering a beautiful pattern of concentric circles of overlapping scales. The **male** (fig. 1c) is also humpless, but it is a little squarer in outline behind than the female, from the ventral surface having been produced into a sheath for the penis.[1]

Length. Female, $\frac{1}{24}$ inch; male, $\frac{1}{48}$ inch. **Habitat.** Ponds and ditches in many parts of England: not uncommon.

[1] In 1871 I found an *Asplanchna* apparently not distinguishable from *A. Brightwellii*, and whose male (*Mon. Micr. J.* pl. xci.) had only two lateral humps. Mr. T. Bolton has lately found both sexes near Birmingham. I have named it provisionally *A. intermedia.*

A. PRIODONTA, *Gosse.*

(Pl. XII. fig. 2.)

Aspl. nelæa priodonta [1] . . Gosse, *Ann. Nat. Hist.* 2 Ser. vol. vi. 1850, p. 18, pls. i. ii.

SP. CII. **Female** *without humps;* **eyes** three; **rami** *broadening to the free ends, their inner edges serrated;* **contractile vesicle** *about equal to the two gastric glands together;* **vibratile tags** *four on each side, attached to a single coil of tubes;* **ovary** *roundish;* **male** *without humps.*

Mr. Dalrymple's exhaustive memoir on *A. Brightwellii* was followed soon afterwards by a similar paper of Mr. Gosse's on his new species, *A. priodonta,* in which its structure is described with the greatest minuteness and care. *A. priodonta* was discovered by Mr. Gosse in the Serpentine, in Hyde Park, in 1850. It is much smaller than *A. Brightwellii,* its outline more oval, its head more conical. The jaws are broader: and instead of a projecting tooth on the inner side, they have this edge minutely serrated, with the tip forming two curved long teeth (figs. 2*f*, 2*g*). Each carries a spine proceeding from its back and curving round nearly parallel to its extremity. The gastric glands are situated on the œsophagus itself, a little above the stomach, not on it. But the chief differences lie in the vascular and reproductive systems. The contractile vesicle, when full, is globular and small, being scarcely, if at all, bigger than the two gastric glands together. The flocculent ribbons that support the lateral canals have their middle parts wrinkled into a large coil of four or five pairs of doublings, and on this coil are placed four vibratile tags ; so that there are only eight tags in all. The **ovary** is roundish and very small ; it is shown in fig. 2*a* with its germs, and the ovisac wrinkled up close to it. In fig. 2 it is behind and above the ovisac, which contains a developing ovum with salmon-coloured oil-globules in it. There are three crimson **eyes** (fig. 2*b*), one under the centre of the corona and one on each side of it ; each resting on a **nervous ganglion.**

The **male** (fig. 2*c*) differs hardly at all in its internal structure from that of *A. Ebbesbornii,* though very different in shape. Its **sperm-sac** is supported by a strip of tissue that hangs from the head, and resembles in shape and position the alimentary canal of the female. It is, however, imperforate and structureless, and seems to have no other office than to support the sperm-sac and penis. The **nervous ganglion** (fig. 2*d*) is unusually conspicuous. Two of its four diverging threads pass downwards to the dorso-lateral rocket-headed antennæ (figs. 2*c*, 2*e*), and two pass upwards to similar antennæ on the two apices of the corona.

By slightly compressing a male, I put beyond question the fact that the contractile vesicle empties itself *outward* through the cloaca ; for under slight pressure the vesicle contracted slowly, by stages as it were, collapsing partially in separate efforts instead of closing at once. As it did so, I distinctly saw, at each effort, the gradual passage of a plug of fluid down the cloaca, dilating its walls as it went.

Length. Female, $\frac{1}{32}$ inch ; male, $\frac{1}{72}$ inch. **Habitat.** Kensington Gardens, Serpentine (P.H.G.) ; ponds and ditches round Clifton and Birmingham (C.T.H. ; T.B.) : not uncommon.

---

[1] Ehrenberg's *Notommata syrinx* is wonderfully like *A. priodonta,* but is said to have a cloaca, and a minute foot and toes. His *N. myrmeleo* is unknown in England, but Leydig has made it clear (*Ueb. d. Bau. d. Räderth.* p. 20, Taf. iv. fig. 36) that in this instance Ehrenberg has made a mistake, and that the Rotiferon has not got the cloaca which Ehrenberg describes. It is therefore an *Asplanchna* with a foot ; one much resembling that of *Notops clavulata.* Its jaws, ovary, vascular system and eye resemble those of *A. Brightwellii.*

Leydig (*loc. cit.*) has described another *Notommata, N. Siebőldii,* which is a true *Asplanchna.* The female closely resembles *A. Brightwellii* ; but the male, which is conical in shape, has four humps, two lateral ones and two on the neck, just like those of the male of *A. Ebbesbornii.*

Genus SACCULUS. *Gosse.*

GEN. CII. **Corona** *the one apex;* **trophi** *inclosed in a mastax, ramate, with unequal mallei, very evanescent;* **alimentary canal** *very large, having eight coeci,* **eggs** *attached after deposition.*

S. VIRIDIS, *Gosse.*

(Pl. XI. fig. 2.)

Sacculus viridis . .	Gosse, *Ann. Nat. Hist.* 2 Ser. vol. viii. 1851. p. 198.
Ascomorpha helvetica .	Perty, *Zur Kenntniss kleinst. Lebensf.* 1852. p. 39.
Ascomorpha germanica (?) .	Leydig, *Ueb. d. Bau. d. Räderth.* 1851. p. 11, Taf. iii. fig. 34.
Sacculus viridis . . .	Gosse, *Phil. Trans.* 1857, p. 320, pl. xv, figs. 21–26.
Ascomorpha saltans (?) . .	Bartsch, *Rot. Hungariae,* 1877, appendix, Tab. ii. fig. 17.

This beautiful " little green sac " was first discovered by Mr. Gosse, in the summer of 1850, in a small pool on Hampstead Heath, and was described by him in the "Annals of Natural History" in 1851. It haunts the bottoms of small pools on heaths and in plantations, and I have occasionally found it roaming over the vegetable sediment at the shallow ends of cattle-ponds. It is not a common creature, and it is an extremely difficult one to study, for its skin is thick and rough, and its huge stomach hides almost all the other organs.

The **ciliary wreath** is a simple ring of cilia with three or four large styles set in it at intervals. Just below the wreath, on the dorsal side, is a comparatively clear space of external surface (figs. 2, 2b), which is shagreened, as it were, with diamond-shaped clusters of granules. Through this can be seen the **nervous ganglion**, bearing a crimson eye, and I think the front portions of two spherical **gastric glands**. The ovoid **mastax** can be readily seen in the side view; it contains delicate triangular rami on a long slender fulcrum and two rod-like mallei, which Mr. Gosse thinks are of unequal length. The shape of the **stomach** is very unusual. Seen dorsally it appears to consist of two cylindrical sacs, one on each side, tapering in front, curved towards each other behind, and connected by a broad cross sac, so as to have a rude resemblance to a letter H. A lateral view shows four short, equal sacs hanging down from the long side sacs ; two on each. The whole of this strange stomach is stuffed full of beautiful green oval bodies, which are probably the zoospores of algæ. At each of the four junctions of the short sacs with the long ones there is a dark-brown spot, which seems to consist of decomposing food. No observer has seen any fæcal discharge, or detected a cloaca ; and although I made several specimens fast for some hours in clear water I failed to obtain one with a comparatively empty stomach. The **ovary** lies between the four short cæcal prolongations of the stomach, and the **contractile vesicle** lies deep down below the connecting cross sac. The animal carries its **eggs** attached to its hind extremity ; bearing sometimes one or two female eggs, or occasionally a whole cluster of small male eggs.

From one of the eggs, which, before maturity, are much clouded and spotted with granules and globules, a young one was produced in my live-box, which was, I doubt not, a **male** (fig. 2c). I could not detect any **eye** (though this organ is conspicuous in the parent) nor any internal organisation ; nothing but a confused assemblage of granules and globules ; even the ordinary opaque masses were not present. The form somewhat resembled that of an amphora with a short wide neck ; the frontal cilia were very large, but the motion was not rapid, nor was the animal wild, as male Rotifera usually are.—P.H.G. ("Phil. Trans." loc. cit.).

Length. Female, $\frac{1}{30}$ inch ; male, $\frac{1}{120}$ inch. **Habitat.** Hampstead Heath (P.H.G.) ; Clifton (C.T.H.).

Family VII. SYNCHÆTADÆ.

Corona *a transverse spheroidal segment, sometimes much flattened, with styligerous prominences;* **ciliary wreath** *a single interrupted or continuous marginal curve, encircling the corona;* **mas'ax** *very large, pear-shaped;* **trophi** *forcipate;* **foot** *minute, furcate.*

Genus SYNCHÆTA.

GEN. CH. Form *usually that of a long cone whose apex is the foot; front furnished with two ciliated* **auricles;** **ciliary wreath** *of interrupted curves;* **foot** *minute, furcate.*

Though this genus consists of only a few species yet it is alike interesting in its structure and its habits. The various species differ from each other chiefly in the shape of the body and of the coronal head, as well as in the number and position of the tactile organs. There is, too, at least one species which is marine, and which has been said by its presence to render sea-water luminous.

In all the species the striking characters are the swift and varied motions, the ciliated auricles, the huge mastax, and the long delicate œsophagus closely resembling that of *Asplanchna.*

S. PECTINATA, *Ehrenberg.*

(Pl. XIII. fig. 3.)

Synchæta pectinata	. .	Ehrenberg, *Die Infus.* 1838, p. 137, Taf. liii. fig. 1.
Synchæta mordax .	. .	Gosse, *Ann. Nat. Hist.* 2 Ser. vol. viii. 1851, p. 200.
Synchæta pectinata	. .	Leydig, *Ueb. d. Bau d. Räderth.* 1854, p. 41.
Synchæta mordax .	. .	Pritchard, *Infusoria,* 1861, p. 686, pl. xxxiii. fig. 122.
„ „ .	. .	Hudson, *Mon. Mier. J.* vol. iv. 1870, p. 26, pl. lvi.

SP. CH. Body *a swollen cone;* **auricles** *very long, pointed, usually pendent;* **coronal head** *very large and convex; two club-shaped* **prominences** *in front, each crowned with a wide brush of setæ; four* **styles***, the outer pair sometimes compound.*

This is the finest and most vigorous of the *Synchætæ.* No one can watch it swimming in ample space, without marvelling at the energy of this living speck, and admiring the grace and ease of its varied motions. No swift is more untiring in its flight. Now it sweeps along in spiral turns from the surface to the bottom, and now it darts through the green branches of the water crow-foot to hang motionless over a leaf like a hovering fly in summer; motionless, and yet with its front all ringed with the halo of its furiously lashing cilia. The **auricles,** which seem mere rudimentary stumps, are really most effective organs of locomotion. They are tongue-shaped fleshy protuberances, edged with powerful cilia; and, as they can be set by special muscles at various angles to the body, the creature can dart, wheel, and stop, with the greatest ease. The **trunk,** seen ventrally, appears to be a cone tapering to a small foot divided into two minute toes, but the side view shows the dorsal surface rising behind the head into a distinct hump. The **coronal head** is nearly half a spheroid. Round its base on the dorsal side are four semicircular curves of small cilia, and two similar curves edge cup-like protuberances on the ventral side. These cups can be lowered or raised a little at will, and their contour altered so as to enable their fringe of cilia to sweep the food effectively between the two into the **buccal orifice.** This lies near the top of the coronal head towards the ventral side. It can be seen only by looking directly at the top of the head; no dorsal or side view will show it, and, unluckily, these are the only views that a compressorium will yield. I have, however, on one or two occasions found a *Synchæta,* left in an open cell, swimming feebly in an upright position just before it died; and I have thus caught sight of the buccal orifice. It is an oval opening between the ventral cups, and overshadowed by two projections each bearing a fan of styles. As the atoms of food are swept towards the orifice, the fans are bent over it, and the styles lash the water to drive downwards any escaping prey. Many Rotifera have a similar contrivance, notably the *Brachioni,* whose coronal styles form quite a dome over the buccal orifice. The corona bears also four **tactile organs,** two towards the dorsal, and two towards the ventral side; and each consisting of one or more styles issuing from a small prominence, and set in a short cylinder. To the end of the cylinder a muscle is attached, so that by this means the styles it carries can be withdrawn nearly below the surface. There is yet another organ of touch. At the summit of the dorsal hump meet two rocket-headed **antennæ,** each bearing a tuft of setæ; and the two tufts issue together from the same opening. A nerve-thread passes from each rocket-head to a **nervous ganglion** in the head, on which is seated a bluish-purple eye-spot [1] bearing a refractive body, shaped somewhat like a truncated cone.

The great **mastax** is a pear-shaped body placed so close to the buccal orifice that there is no room for a buccal funnel. The forcipate trophi are driven snapping through the buccal orifice by the great striated muscles shown in fig. 3. A V-shaped one embraces the end of the fulcrum, and one on each side acts on each malleus. The **œsophagus** closely resembles that of *Asplanchna*, and like that is often used as a supplementary stomach. Indeed, I have seen it so full that, for a moment, I did not recognize *Synchæta*, and thought that I had found a new Rotiferon. The gastric glands are small and round; the foot glands obvious. The **stomach** is usually yellow, but sometimes brown; more rarely, pink. It is round, with thick walls formed of very large cells. Sometimes these are studded all over with oil-globules, as in fig. 3. The **lateral canals** with their vibratile tags can be plainly seen in the lower half of the trunk, but do not appear to run up to the head in the usual fashion. There is a small round contractile vesicle just above the foot. The longitudinal and transverse **muscles**, as well as those that work the auricles, can be seen in fig. 3, and need no further description.

The **male is** as yet unknown.

Length, $\frac{1}{70}$ inch. **Habitat.** Clear ponds and reservoirs: common.

SYNCHÆTA BALTICA, *Ehrenberg.*

(Pl. XIII. fig. 1.)

Synchæta baltica		Ehrenberg, *Die Infus.* p. 137, Taf. liii. fig. 5.
" "		Gosse, *Tenby,* p. 274, pl. xiv.

[SP. CH. Body *cylindric, becoming conical behind, or bell-shaped: gibbous dorsally;* **rotatory clusters** *four;* **styles** *four;* **crest** *single, sessile. Luminous. Marine.*

The **form** is usually that of a bell, or a long cone, viewed dorsally; but much arched, viewed laterally. The **auricular lobes** are very small: two equidistant setæ radiate from the occiput. There are two ciliated eminences besides the auricles, and a medial crest, smooth-edged. The front is rather wider than the body, whose hinder part tapers to a well-jointed foot, and two very minute conical **toes**. Conspicuous pear-shaped **mucus-glands** from these run up through the foot.

A large **red eye** is seated at the end of a cylindrical brain-sac: below which is the vast **mastax** of normal structure. We see, now and then, a momentary snatching action with this organ, of which I have not been able to define the actual seat. A sudden trembling also occasionally passes through the whole fore parts. A long œsophagus leads to a small sacculate yellow **stomach**, on which are seated ample **gastric glands**. A small intestine opens into the cloaca, above which projects a wart, which is a true tail. A band-like **ovary**, of horseshoe form, the ends forward, lies in the lower belly; and behind this a small **contractile bladder**, whose period is $2\frac{1}{2}$ times a minute. The **muscular system** is very distinct: a cord from the frontal region is inserted in the skin of the back, thrown into sigmoid curves during contraction; other cords proceed from the hind-head to various points near the middle of the trunk; one (pair?) proceeds from the mid-back to the side: five or six bands run transversely across the back. There are, moreover, long diagonal bands down the sides.

The **brilliant translucency of** this animal makes it a very charming object, particularly when well illuminated on a dark ground, when the eye shines out like a ruby, and the whole body resembles a sparkling diamond. Its movements are vivacious and elegant. It shoots rapidly along, or circles about in giddy dance, in company with its fellows, sometimes near the surface, sometimes just over the bottom of its prison. Occasionally the foot and tiny toes are drawn up into the body, and then suddenly thrust down, and bent up from side to side, as a dog wags his tail. Sometimes the two ear-

[1] Crimson by lamp-light and dark field illumination.

lobes are brought forward, and then spasmodically spring back to their ordinary position, when the creature shoots forward with redoubled energy. All its actions display vigour and precision ; and convey the impression of intelligence and will.

The most interesting fact connected with its history is that it seems to be one of the sources of the phosphoric light which often pervades the waves of the sea. In July, 1854, at Tenby, I saw the water within the harbour splendidly luminous. No trace of light, indeed, appeared on the smooth surface, but when this was agitated it blazed. The finest effect was produced by dashing a large stone down from the quay : every spray that splashed up was luminous ; and thus a momentary star of many irregular rays of light was made, some of the lines reaching to fifteen or twenty feet. At the same moment a great circular wave was raised, which took the appearance of a bank, or annular agger, most intensely lustrous, but so transient that the progression of the wave could not be traced ; the light sank into darkness in an instant. The Bristol steamer was just leaving the wharf, and an impatient stroke or two from her paddles illuminated the dark water under her quarter, and the lowest step of the quay stairs was every instant covered with sparks, like diamond dust, by the tiny wavelets that washed over it. On examination, I found specimens of *S. Baltica* in it; associated, however, with other animalcules, both larger and smaller, which were indubitably luminous, as *Noctiluca* and *Ceratium*.

I first met with this species in July, 1850, in water from the mouth of the Naze, in Essex. Mr. Hood has lately found it in the estuary of the Tay, in Scotland, with many other marine Rotifera ; and has communicated living specimens to me, one of which has contributed to the present description.—P.H.G.]

Length, $\frac{1}{15}$ to $\frac{1}{15}$ inch ; width, $\frac{1}{33}$ to $\frac{1}{50}$ inch. **Habitat.** Sea-water ; coasts of England, Wales, and Scotland (P.H.G. ; J.H.).

S. OBLONGA ?, *Ehrenberg.*
(Pl. XIII. fig. 4.)

Synchæta oblonga (?) . . . . Ehrenberg, *Die Infus.* 1838, p. 438, Taf. liii. fig. 6.

[SP. CII. **Body** *ovate or pyriform ;* **head** *very large ;* **auricles** *wide ;* " rotatory clusters *six ;* **styles** *four ;* **crest** *single, sessile* "; **toe** *single, minute, without foot. Lacustrine.*

A species which I met with in the Watering Pond on Hampstead Heath in 1850, I concluded to be *S. oblonga* of Ehrenberg, though I could not identify all the characters. But a single specimen occurred, and I have never seen it since, till in November 1885, in a tube dipped from Keeper's Pond, Birmingham, and sent to me by Mr. Bolton, I met with a second example, recently dead, but in fair preservation.

The **front** has two minute ridges, each with an edging of short comb-like spines ; outside these are the two knobbed **antennæ**, on which I did not detect any brushes of divergent setæ. A good deal lower, on the slope of the auricle, on one side, was a long stiff bristle, doubtless one of a pair. The auricles are very large, and each is pervaded by a chain of globose bodies, possibly ganglia, which, having passed around the swollen extremity, turns back at least as far as the base of the bristle. A vast **mastax** exists, whose chief visible feature is a stout incus, whose wide rami appear as diverging lines. A very long, delicate, corrugated œsophagus leads to a small, globose, sacculate **stomach** (which recalls the structure common in the *Asplanchnæ*) filled with green food ; thence a thick, much-wrinkled intestine passes straight to the extremity, where is a very minute, conical **toe**, which I could not by any effort divide. A glandular thread runs from the tip to a minute globose vacuole (?) at its base. Three great **ova**, colourless but turbid, were in the body-cavity, from the appearance of which I should conjecture the animal to be viviparous. Various muscles and nervous (?) threads are shown in the figure.—P.H.G.]

Length. $\frac{1}{8}$ inch ; greatest width, $\frac{1}{13}$ inch. **Habitat.** Hampstead ; Birmingham (P.H.G.) : rare.

S. TREMULA.

(Pl. XIII. fig. 2.)

Synchæta tremula .	Ehrenberg, *Die Infus.* 1838, p. 138, Taf. lii. fig. 7.
„ „ .	Leydig, *Ueb. d. Bau d. Räderth.* 1854, p. 41.
„ „ .	Pritchard, *Infusoria*, 1861, p. 686.

SP. CH. **Body** *a slender cone ;* **coronal head** *nearly truncate ;* **auricles** *scarcely protuberant ;* **setæ** *four ;* **no club-shaped prominences** *; a sudden diminution in girth below the cloaca.*

S. tremula is rather smaller than *S. pectinata*, and its habits are different. It loves to twirl round its own longer axis at the end of a thread stretching from its toes ; and, so twisting, to drift lazily along with the current which bears the object to which it is attached. Its **coronal head** is almost flat, and the side auricles are nearly in the same plane with it. This makes the animal strikingly unlike *S. pectinata* in outline. It has no crests on its corona ; only four long curved styles, similar to those of *S. pectinata.* Its **stomach** is generally full of a rich brown food, and I have sometimes captured specimens with the œsophagus at the same time stuffed with some pinkish substance. Its **eye** is an intensely dark-red, and Mr. Gosse has detected a refractive body imbedded in the pigment.[1] There is a rocket-shaped **antenna** (fig. 2b) on each side of the trunk just above the foot ; organs that I have failed to detect in *S. pectinata.* In all other respects the structure of the two species is almost identical.

[In one of the shallow evaporating tanks in my orchid house, I found (at the end of May) this pretty species swarming. It plays, by myriads, just above the dull-green flocose sediment that settles on the bottom. I learn, from this colony, a habit which I think has not been recognised as proper to this genus—viz. that, like the *Brachioni* and *Anuræ*, and one or two other genera, *Synchæta* retains its **egg** after discharge, attached to its own body, just behind the foot. The egg, which I saw, was nearly globular, of a pale yellow hue, granular by the process of segmentation.—P.H.G.]

I found the **male** (fig. 2c) in the winter of 1870. It is much smaller than the female, narrower for its length, but otherwise much like her in shape, and with the same four styles on the coronal head. I distinctly noticed in it the entire absence of the nutritive system ; but its irrepressible energy prevented me from obtaining more than a fleeting view of the sperm-sac and penis.

Length, $\frac{1}{130}$ inch. **Habitat.** Clear ponds : common.

---

[1] On the occipital aspect of the brain-mass is seated an eye-spot, always conspicuous both from its great size and from its intense colour, a red so deep as to be practically black. Its outline varies much ; but normally it is a hemisphere, or rounded cone ; often it seems homogeneous, but occasionally we see that it is composed of a multitude of pigment cells agglomerated together and inclosed within a transparent capsule, whose walls I have frequently detected of a thickness greater than that of one of the pigment cells. But more than this ; I have seen, so often as to have no doubt of its presence, an ovate transparent cell, let-in, as it were, into the coloured body of the eye, the dark pigment rising on each side so as to embrace the base of it. I venture to think this a crystalline lens. - P.H.G.]

PLATE XIII.

PLATE XIV.

PLATE XV.

www.ingramcontent.com/pod-product-compliance
Lightning Source LLC
Chambersburg PA
CBHW021704210326
41599CB00013B/1516